Fire Protection Approaches in Site Plan Review

Fire Protection Approaches in Site Plan Review

Mohammad Nabeel Waseem

CRC Press
Taylor & Francis Group
Boca Raton London New York

CRC Press is an imprint of the
Taylor & Francis Group, an **informa** business

CRC Press
2385 NW Executive Center Drive, Suite 320, Boca Raton FL 33431

and by CRC Press
4 Park Square, Milton Park, Abingdon, Oxon, OX14 4RN

First issued in hardback 2019

CRC Press is an imprint of Taylor & Francis Group, LLC

ISBN-13: 978-1-4987-4178-1 (hbk)

Visit the Taylor & Francis Web site at
http://www.taylorandfrancis.com

and the CRC Press Web site at
http://www.crcpress.com

To my parents (Mohammad and Saeeda) for raising me and teaching me everything I know

To my wife Amber for her love and support

To all plan reviewers and engineers; those who sit behind the desk going unnoticed, spending their lives serving and protecting the people of this world

Contents

Preface ix
Acknowledgments xi
Author xiii
Introduction xv
Analysis xix

Chapter 1 Site Plan Basics 1

Building Address 4
Measuring Scales 7
Profile 8
Existing Condition/Demolition 8
Revisions 9
Site Renovation 9
Local Fire Alarm Notification 9
Record Keeping 9
References 10

Chapter 2 Grading 11

Public versus Private Streets 14
References 15

Chapter 3 Fire Flow 17

ISU Method 17
Illinois Institute of Technology Research Institute Method 18
ISO Method 18
Fairfax County Method 18
Occupancy Reduction—"OR" Value 19
Exposure Distance—"ED" Value 20
IFC Method 21
Minimum and Maximum Fire Flows 21
Example of the Fairfax County Method 21
References 24

Chapter 4 Fire Hydrants 25

Dry Hydrants 25
Wet Hydrants 25
Flow and Pressure at Hydrants 27
Clear Space around Fire Hydrants 31

Distance between Fire Hydrants 32
Fire Hydrant Availability and Access 35
References 39

Chapter 5 Underground Fire Lines 41

Tall Buildings 42
Case Study: NFPA 13R Sprinkler System 45
Profiles 46
Easement 49
Site Renovations 49
Case Study 49
Reference 50

Chapter 6 Fire Department Connections 51

Plan View 53
Fire Hydrant Serving A Siamese Connection 56
FDCs for High-Rise Buildings 57
Case Study 59
Pumper Test Header 60
Free-Standing FDC 61
References 62

Chapter 7 Fire Truck Access 63

Dead Ends 67
Fire Lane Signage 70
References 71

Chapter 8 Aerial Ladder Truck Access 73

How Much Access Is Enough Access? 81
References 85

Chapter 9 Code Modifications 87

Chapter 10 Conclusion 91

Appendix 93

Glossary 99

Index 101

Preface

In today's fast-paced transforming industry of uniquely designed buildings with integrated security features, there need to be ground rules for evaluating site plans with regard to fire safety and emergency operations. The concept of square buildings with all-around access and approach has long faded. Zoning requirements, topography challenges, green design, and the need to maximize tenable space per square foot of area have led to complex urbanized layouts. This book provides guidelines for fire site plan review. It provides a technical examination of how to review site plans, taking into consideration an engineering approach to fire emergency service, codes and standards requirements, and best practices. The material mentioned in this text can be amended at the local level so that it benefits the entire organization. The overall objective is to bring uniformity; to develop a procedure so that the men and women responding to the emergency can do their jobs efficiently and safely.

Whether you are just starting to grasp the concepts of site plans or you are an experienced reviewer or an engineer looking for help in getting approval from the authority having jurisdiction (AHJ), this book will provide a fundamental knowledge of fire principles related to site review. It lays out the procedures to consider when examining site plans relating to fire protection features.

Chapter 1 provides an overview of site plans, discussing vital items that should be on a site plan for fire analysis, what kind of information needs to be evaluated, and the items a reviewer should be familiar with as he/she begins to undertake a review.

Chapter 2 establishes the importance of the roads used by fire emergency personnel to get to a site and for a set up, with specific reference to grading. Having an optimal area for a set up will not only provide an accessible route to get to the building, but will also decrease response time.

Chapter 3 dives into the water requirement for a fully engulfed building. It offers techniques to evaluate the quantity of water that is needed to put out fire in a building. The term *fire flow* is introduced.

Chapter 4 discusses the strategic positioning of fire hydrants with reference to the building and to roads. The two different types of fire hydrants, wet and dry, are examined, along with the sizing of adequate feeds to them and the obtainable fire flow.

Chapter 5 examines underground fire service lines that supply water to the active fire protection system of the building (e.g., the sprinkler system). Little attention has been given to the examination of fire lines, and acceptance/inspection is left to field personnel. Fire lines are perhaps the most important single element of the building's automatic fire protection system. Failure of the fire line results in an unprotected building.

Chapter 6 highlights Siamese connections, also known as fire department connections (FDC). For most automatic systems, like the automatic sprinkler system, they are used as a supplementary means to aid the fixed fire protection/suppression system, but for standpipes they are the primary means to supply pressure and flow (e.g., manual standpipes) for internal handheld-hose firefighting. Location and positioning of FDC, clearance from obstructions, size, and labeling are all looked at in this chapter.

Chapter 7 covers access for emergency vehicles departing from the fire station and en route to the building. The discussion involves having adequate entry points to the site, as well as leaving the site after the incident, with specific attention to dead ends, turnarounds, and street widths.

Chapter 8 proceeds to the analysis of an advanced type of truck, the aerial ladder truck. When are these trucks required, what changes have to be made to the site to support them, and what is the purpose of these units are discussed.

Chapter 9 concludes with the concept of code modifications. These can be thought of as alternatives to the code but are not waivers; rather, their intent is to provide equivalence so that the intent and spirit of the code are met.

The fire protection and life safety issues addressed in this book are discussed on the basis of emergency access to the site and exterior fire protection features. These are the fundamental elements of the fire site plan. While every site should be reviewed for fire protection, this book tends to focus more on urban rather than rural communities. The components of a fire review not only affect internal fire and life safety features, but at times can also prompt a significant change to the architectural design of the building. The codes and standards referenced in this book are the latest available at the time this book was written. These include the International Building Code (IBC), the International Fire Code (IFC), and the National Fire Protection Association (NFPA) standards.

Unit conversion
1 foot (ft) = 0.3048 meters (m)
1 foot = 12 inches
1 inch = 2.54 centimeters (cm) = 25.4 millimeters (mm)
1 gallon = 3.785 liters
1 gallons per minute (gpm) = 3.785 liters per minute (lpm)
feet^2 = square feet or sqft representing area

Acknowledgments

This book could not have been written without the help and guidance of the following personnel:

Adeel Waseem
Maurice Jones
Sandra Ward
Eric Forbach
Jeffrey Allen
Shamsher (Sam) Singh
Mike Long
Ronald Klus
Richard Twomey
Sania Waseem
John Walser
Cheryl Wood

To my colleagues at Fairfax County, Fire Prevention Division, Fire Plan Review Department for providing guidance, and especially to the person who taught me how to review site plans, David J. Thomas, MSCE, PhD, PE, many thanks.

Author

Mohammad Nabeel Waseem Hasan has an Advanced Diploma from the Seneca College of Applied Arts and Technology in fire protection engineering technology (Ontario, Canada), a Bachelor of Science in fire safety and engineering technology from the University of Cincinnati, and a Master of Engineering in fire protection engineering from the University of Maryland. In addition, he holds a Master of Business Administration (concentration in project management) from Columbia Southern University, where he also served as a part-time adjunct faculty member, teaching various fire science courses. He is a professional member of the Society of Fire Protection Engineers and a certified fire protection specialist. He holds a National Institute for Certification in Engineering Technologies Level III certification in water-based systems, and is a certified fire plans examiner, a fire inspector, and a licensed fire protection engineer in the state of Maryland and the Commonwealth of Virginia. He has over a decade of experience in the fire protection industry, and served as the lead fire protection site reviewer for the Office of the Fire Marshal in Fairfax County, Virginia.

FEEDBACK

The author welcomes any suggestions to improve the material in this text. Comments and questions can be directed to the author via email at flamenab@hotmail.com with the subject line "Book: Fire Protection Approaches in Site Plan Review."

Introduction

The significance of a site plan may not be easily understood. Information can be contained in a single sheet or can encompass dozens or even hundreds of sheets. Little attention is given to site plans by individuals and agencies, and few people understand their importance in terms of safety.

Let us begin by answering the question: What is a site plan? A site plan is a detailed engineering drawing of proposed improvements to a given lot. It is a written and visual representation of any elements that change the site from its existing condition, including developing a new building or demolishing an old one, building expansions or additions, making changes to grading, landscape, utilities (e.g., water lines, overhead power lines, generators, underground tanks, and storm water), roads, and fire lanes.

When is a site plan required and when does it have to be submitted? Every jurisdiction has its own requirements. Contact the local building department and find out when one is required. Often, changes to a lot are not significant and a minimal submission is needed—a minor site plan. It is good practice to obtain this information well in advance so that you are aware of which agencies are involved and what their requirements are. A site plan can also be requested by other agencies, including state and national organizations. For example, jurisdictions that have adopted the International Building Code (IBC) can request a site plan.

A site plan is not just used for checking soil conditions, setbacks, utilities, roads, storm water, water, landscape, zoning requirements, and parcels. It is crucial for fire safety analysis. As the building is designed and takes shape, site plans must be thoroughly reviewed for code compliance with all regulations. Agencies such as zoning, land development, etc. review specific items against the codes, county policies, standards, and regulations. This book will help you see a site plan through the eyes of a fire protection plan examiner.

When should a site plan be reviewed? Whether you are in a rural area or in an urbanized setting, every development goes through planning. This is to allow for coordination, to achieve a feasible design, and to identify problems so that the project can be cost efficient with the least amount of obstacles. There are many examples where poor design and planning combined with inadequate site plan reviews have led to delays in project completion and increased cost. For example, access is insufficient for emergency responders to get to the building; Siamese connections are not labeled or are on the wrong side of the building; or fire hydrants are randomly placed miles away from a house on fire. Entire buildings have been destroyed with millions of dollars in damages due to these errors.

Consider the following case:

> On the morning of April 1, 2014, in Rockville, Maryland, despite more than 200 firefighters responding to an apartment fire involving 150 brand new units, a loss of US$20 million occurred. The entire building was destroyed as the fire, aided by the wind, spread across 3 acres of the complex. The apartments were nearly complete, with tenants ready to move in. The building had a fully automatic fire suppression system, but

it was not yet operational. Crews had to battle the fire using heavy master streams and aerial ladder trucks to prevent the spread of fire to adjacent occupied buildings, which included the Substance Abuse and Mental Health Services Administration (Figures 0.1 through 0.3). Challenges were encountered in getting access to the hot spots and in attacking the front of the fire, which had rapidly spread to the rear of the building. Fire truck access had not been provided to the rear of the building, so water could not be applied directly. The aerial ladder trucks had to be set up on an interstate highway

FIGURE 0.1 Firefighters use an aerial ladder truck to project a stream of water to control the fire. (Photo by Peter Piringer. 2014. Massive Fire Engulfs, Damages Montgomery County Apartment Building, Retrieved September 13, 2014 from http://wtop.com/news/2014/04/massive-fire-engulfs-damages-montgomery-county-apartment-building-video/slide/2/. With permission.)

FIGURE 0.2 Destroyed apartment complex. (Photo by Dave Thomas. With permission.)

FIGURE 0.3 Roof collapse. (Photo by Dave Thomas. With permission.)

> ramp to achieve an appropriate angle of attack. Many of the neighborhood residents were trapped in their community, unable to get out due to hoses blocking the streets. Road closures caused heavy traffic delays and congestion as equipment was moved to battle the blaze. (Bell and Gonzales, 2014)

Whether it is an urban environment or a rural community setting, a detailed analysis of site plan review is needed now more than ever in the twenty-first century and beyond. Fire site plan concepts must be not only in the minds of civil engineers, but also in the mind of the architect as the conceptual design takes place. It is vital for the fire plan review and fire site analysis to occur ahead of time with the planning and zoning commission. This will avoid many problems in design changes further down the path, and help to fit in the fire features from the beginning.

REFERENCE

Bell, B., and Gonzales, J. 2014. Rockville three-alarm fire burns unfinished apartment complex. Allbritton Communications Company (ABC), retrieved September 13, 2014 from http://www.wjla.com/articles/2014/04/rockville-two-alarm-fire-burning-unfinished-apartment-complex-101717.html

Analysis

Issues pertaining to fire protection on a site plan should not be an afterthought or the responsibility of the fire plan reviewers. The concepts discussed in the chapters of this book must be kept in mind by architects/engineers throughout the entire design process. This can reduce, if not eliminate, many issues that will occur during the review process. Overall, it will speed up the approval process and save money by not having expensive changes arise during construction or field inspections, potentially causing delay when the building is about to open. The example mentioned in the Introduction, of the fire that occurred in the United States in 2014 that resulted in the destruction of an entire apartment complex with a US$20 million loss, might not have been prevented, but perhaps with adequate fire truck access to the rear of the building the damage might have been less.

Many developers, owners, and contractors think that their building will never catch fire or burn because their building has an automatic sprinkler system. The apartment building above was fully sprinklered, but during construction the system was not yet operational; nor was it required to be until construction was complete.

It can be argued that if the fire suppression sprinkler system had been active, the fire would not have spread. Sprinklers provide interior protection, and it is true that a "sprinkler system is considered reliable and effective when properly designed, installed and maintained" (Long et al., 2010). The NFPA states "that a properly installed and maintained automatic sprinkler system will reduce the average property loss from fire by one-half to two-thirds. According to the U.S. Fire Administration (USFA), property losses are 85 percent less in buildings that are protected with fire sprinklers compared to those without sprinklers" (International Fire Service Training Association, 2010, p. 16). No one better than the fire plan reviewer and firefighter understands the importance of sprinklers and how they are one of the greatest inventions. An automatic sprinkler system is invaluable; however, it does not replace firefighters. One key thing to realize is that typical fire sprinklers for office, commercial, mercantile, and school buildings are designed to control the fire, not suppress it. The activated sprinkler heads work by cooling the gas jet layer and preventing fire by preventing flashover conditions. The design is to limit the fire from spreading so that once firefighters respond, they can extinguish it. For certain occupancies, such as warehouses, the design could include the use of early suppression fast response (ESFR) sprinklers, which have the ability to achieve complete suppression. ESFR is not the solution for every occupancy. With them there are additional design factors and increased costs, and the system still does not provide a 100% guarantee. Additionally, sprinklers do little good in an exterior fire aided by strong winds.

At this point it is worth looking at some statistics regarding sprinkler systems and past fires. According to the NFPA Fire Analysis and Research, the report "U.S. Experience with Sprinklers," for fires between 2007 and 2011, indicates the following:

> When sprinklers were present in the fire area of a fire large enough to activate them, in a building not under construction, they operated 91% of the time. When they operated,

they were effective 96% of the time, resulting in a combined performance of operating effectively in 87% of reported fires … (Hall, 2013).

On this basis, there is a 13% chance that sprinkler heads will not be sufficient to control the fire. This may be seen as a low probability, but at what risk? Every effort must be made to ensure life safety and property protection. The counteractive measure against the failure rate comes in the form of manual fire protection, which relies on a fire site plan review to ensure that fire trucks have adequate access, the required fire flow is present, the fire hydrants are correctly located and at the right distance so that water is available, and much more.

Perhaps the article titled "Lessons Learned from Unsatisfactory Sprinkler Performance: An Update on Trends and a Root Cause Discussion from the Investigating Engineer's Perspective" puts it in the best terms: "Oftentimes, performance is affected by factors not linked to the initial design or installation" (Long et al., 2010).

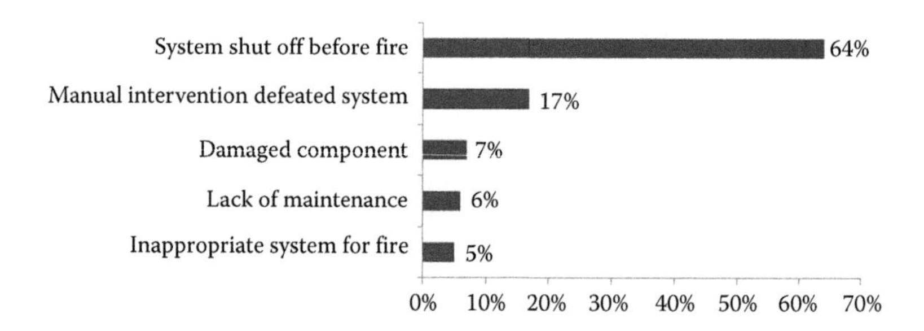

FIGURE 0.4 Reasons sprinklers fail to operate, 2007–2011. (Reprinted from Hall, J.R. 2013. U.S. Experience with Sprinklers, 6/13. National Fire Protection Association. With permission.)

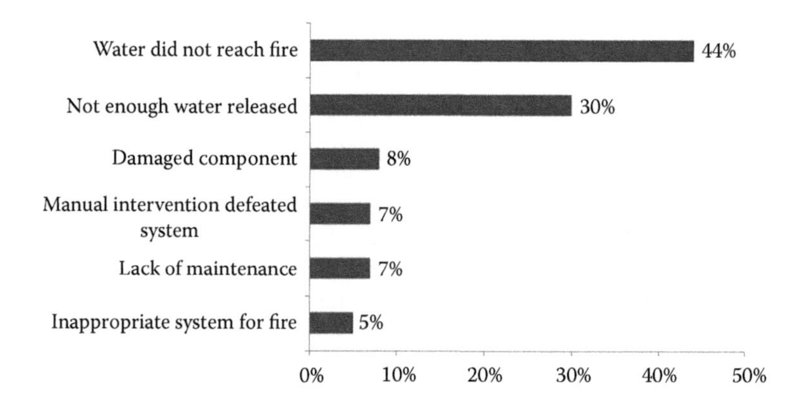

FIGURE 0.5 Reasons sprinklers are ineffective, 2007–2011. (Reprinted from Hall, J.R. 2013. U.S. Experience with Sprinklers, 6/13. National Fire Protection Association. With permission.)

In 2013, the NFPA reported the reasons presented in Figures 0.4 and 0.5 for the failure of sprinklers to operate and for them to be ineffective on fires in the United States during 2007–2011 (Hall, 2013).

Complete fire suppression is dependent on manual intervention. Sprinkler systems are affected by lack of maintenance and testing, failure of main devices due to corrosion or obstruction in supply piping, improper design of the system, failure to control the fire when excessive sprinkler activate beyond the remote area, incorrect installation, change of hazard, obstruction to sprinklers, changes to building interiors, reducing of water supply both in flow and pressure due to development, and exposure to fires from exterior or adjacent buildings.

In 2008, there were 38 fires that resulted in a damage of US$2.34 billion, which killed 15 civilians and injured 60 people, including 32 firefighters (Long et al., 2010). Twenty-one of these fires had a fire suppression system. Consider the following cases of building fires, all of which had automatic sprinkler systems:

- July 2007, Massachusetts, US$26 million loss: Fire occurred in the basement of a fully sprinklered three-story building, 350,000 square feet (sqft) in size, containing fifty-six mercantile businesses.

Reason for fire suppression system failure: Water supply was shut off (Badger, 2008).

- January 2007, Georgia, US$7.5 million loss, one person killed: Fire occurred in a sprinklered machinery room of a three-story building, 245,000 square feet (sqft) in size; seventy-five sprinkler heads operated but were unable to control the fire.

Reason for fire suppression system failure: Improper system maintenance, including sprinkler clearance, and valves not fully opened (Freeman, 2008).

- March 1998, Arizona, US$6 million loss: Merchandise caught fire in a fully sprinklered building with in-rack sprinklers, single-story bulk retail store, 100,000 square feet (sqft) in size; more than 2.5 times the sprinkler heads in remote area activated.

Reason for fire suppression system failure: Change of commodity with sprinkler systems not upgraded for plastics storage (Comeau, 1998).

- November 2008, unknown, US$10 million loss: Fire occurred in a concealed nonsprinklered attic of a sprinklered one/two-story motel building and spread to the sprinklered area of the building in the lobby and guest rooms, where sprinklers were activated but were unable to control the fire.

Reason for fire suppression system failure: NFPA 13R system omits sprinkler protection from attics (Badger, 2008).

All the above buildings had an automatic fire suppression system, yet the fire still spread and caused millions of dollars worth of damage. Additionally, external fires, as a result of faulty wiring, lightning strikes, storage of outdoor combustibles, or exposure from other buildings, cannot be suppressed without manual fire intervention. Combustible exterior walls can play a role in the rapid spread of fire from one floor to another. To quote, my supervisor, Dave Thomas, retired Fire Protection Engineer from Fairfax County, Virginia, "Sprinklers cannot possibly be expected to protect against exterior fire, since exposure sprinklers are mostly not in use and are difficult to implement. This means that the exterior fire spread threat will not ever go away, and that provision for proper access is always required, for any type of structure."

The threat of fire is real. Reflect on the probability of sprinkler suppression success and then the failures, the people injured, and the lives lost. This information is not meant to put a dent on the reliability of sprinklers; rather, it is to point out the facts. A well maintained and properly designed fire suppression system is the best method available to protect against immediate internal fires and to control their spread. With manual firefighting, there can be additional damage due to the excessive hose streams, increasing water damage, and overhaul (the pulling down of walls and ceilings to locate fires), and the lives of the emergency personnel are put at risk. Manual firefighting cannot replace sprinklers, nor can sprinklers replace manual firefighting. Since manual firefighting is the only other suppression method available besides building suppression systems, careful site planning plays a vital role in saving property and lives.

Every community should be analyzed on the basis of its own merits as it is not possible to cover all individual communities. This book will give a general idea of the items to be incorporated in a review.

REFERENCES

Badger, S. 2008. Large-Loss Fires in the United States 2007, National Fire Protection Association, Quincy, MA.

Comeau, E., Duval, R., and Dubay, C. 1998. Bulk Retail Store Fire, Tempe, Arizona, March 19, 1998, National Fire Protection Association, Quincy, MA.

Freeman, P. 2008. Universal Studios Fire: Fire Flow Assessment Report, Los Angeles County Fire Department, Los Angeles.

Hall, J.R. 2013. U.S. Experience with Sprinklers, 6/13. National Fire Protection Association, Quincy, MA. Fire Analysis and Research Division. Retrieved September 14, 2014 from http://www.nfpa.org/~/media/files/research/nfpa%20reports/fire%20protection%20systems/ossprinklers.pdf

International Fire Service Training Association. 2010. *Fire Detection and Suppression Systems.* 4 ed. Stillwater, OK: Fire Protection Publications.

Long, T., Wu, N., and Blum, A. 2010. Lessons learned from unsatisfactory sprinkler performance: An update on trends and a root cause discussion from the investigating engineer's perspective. *Fire Protection Engineering.* Retrieved from http://magazine.sfpe.org/sprinklers/lessons-learned-unsatisfactory-sprinkler-performance-update-trends-and-root-cause-discuss

1 Site Plan Basics

Proper planning of fire protection and life safety features of any site, whether or not an actual building is to be developed, should occur not when the architectural design of the building is complete or when the plans are submitted, but during the planning/zoning review or even at the preliminary developmental stage. The planning/zoning should be attended by the Authority Having Jurisdiction (AHJ). Site plan review is one of the primary steps in evaluating the fire protection features of a building and site. According to the NFPA *Fire Protection Handbook* 20th edition, a site plan "provides an overview of the intended construction in relation to the existing conditions and includes information such as building placement, exposures, size, type of construction, occupancy, water supply from both public and private sources, hydrant placement and access" (NFPA, 2008).

A fire plan reviewer uses the site plan to determine code compliance and to ensure that the site has features that address the needs of emergency response units. The first sheet of any site plan is one of the most important sheets. Usually it is the cover sheet, which contains information for the signing authorities, the deed information, the developer information, tax map number, zoning tabulations, etc. The cover page should also have Fire Marshal (reviewer) information, which should include at least the following information:

- What building code edition the building will be built to, that is, ex. NFPA 5000 2012 or IBC 2009
- Type of construction
- Use group
- Number of stories
- Height of building
- Footprint area (FPA) of the building (i.e., the total allowable area of the largest floor of the building)
- Gross floor area (GFA) of the building (i.e., the total allowable area of the building of all the floors combined)
- Fire flow available in gallons per minute at the residual pressure of 20 psi, and the testing details
- Fire suppression system to be part of the building, that is, whether or not the building has a sprinkler system and the type of sprinkler system if there is one (NFPA 13R, NFPA 13, NFPA 13D, limited area)

This information can be formally included on the site plan cover sheet so that it is quickly available to the reviewer.

Example of an information block specifically for a fire reviewer:

FIRE MARSHAL INFORMATION

Building code in effect and edition: ________________
Type of construction: ________________
Use group: ________________
Number of stories: ________________
Height of building (feet or meter): ________________
Foot print area (feet^2 or meter^2): ________________
Gross floor area (feet^2 or meter^2): ________________
Name of water provider: ________________
Fire hydrant number: ________________
Q20 (gpm or kPa): ________________
Building fully sprinklered: Yes ____ No_____ Partially ______
If yes, type of system:
NFPA 13 _____ NFPA 13R _______
NFPA 13D ______ limited area _______

Including this and other information at the beginning makes it quickly available to the reviewer and provides an overview of the project.

Note that the information required above may not be the same as defined by zoning or other agencies. Zoning may also require similar information, such as height or area, but these may be as per zoning definition and not as per the building/fire code. It is important for the reviewer to be aware of this difference and request for more information if needed. If there are multiple buildings on the site plan, then it may not be feasible to put all the information on the cover sheet. For such cases, a note can be made, for example, "See Sheet 30," to send the reviewer to a sheet that does contain all the information. Many times, engineers consolidate multiple buildings and fire area information into a table. An example is shown below:

Building	Construction Type	Use Group	Height (feet or meter)	Number of Stories	FPA (feet^2 or meter^2)	GFA (feet^2 or meter^2)	Sprinklered
HQ building	IA	B, A3, S2, M (nonseparated mixed use)	200′	10	50,000	450,000	Yes, NFPA 13
Service bay	IIB	S-1, B (separated mixed use)	30′	1 with mezzanine	10,000	12,000	Yes, NFPA 13

Note: Use and construction type per International Building Code (IBC).

During site plan review, for sites that are developing buildings, we are interested in determining the fire areas separated by fire walls, not fire barriers. Only fire walls create separate building as far as building height and area are concerned. Therefore it is important to indicate on the site plan the location of the fire walls and their rating. A reviewer with sufficient knowledge of the building code should verify the rating of the walls to determine if the ratings are in compliance with the code for the use group and the type of building construction, for example, exterior wall rating based on fire separation distance. Fire separation is a distance measured at right angles from the face of a building wall to either the closest interior lot lines, the center line of the street, alley, or public way, or an imaginary line between two buildings on the property (ICC, 2006). Exterior wall ratings must be indicated on the site plan as they affect staging operations and fire access (Figure 1.1).

In the process of site plan review, a fire plan reviewer who has knowledge of the building code can evaluate to a certain limit basic building elements, such as height and area limitations, mixing of use groups, and construction type, while the building design by the architect is taking place. This objective can help mitigate problems than may be encountered later during building plan review.

The cover page of any site plan should also contain a table of contents of all the sheets in the plan set. This could range from a few sheets to well over a hundred. Going through each of these sheets to search for information can be a tedious task and can take hours. A key part of the reviewer's job is developing the skill to be efficient at reviewing a plan, being able to quickly and effectively find the information. For example, erosion and sediment control are irrelevant to a fire reviewer. After going over the Fire Marshal information box on the cover page, the next task is to scan the table of contents and highlight/circle the important sheets. These could be labeled as: details, site plan, grading plan, geometry/layout plan, existing/demo plan, utility plan, water line profile, fire lanes/marshal plan, and landscape plan.

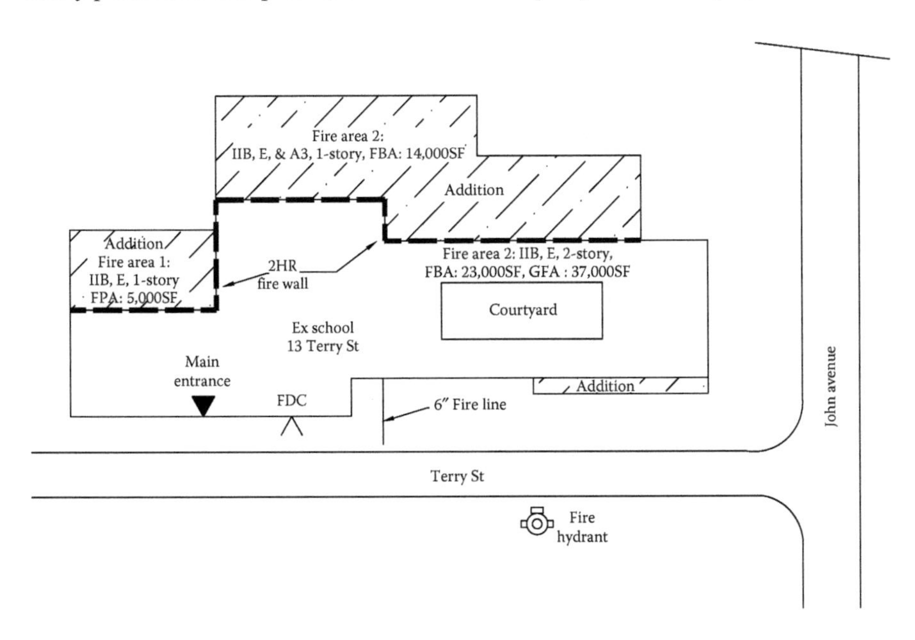

FIGURE 1.1 Example of a school with multiple fire areas separated by fire walls.

The details page should be one relating to water information, such as generic fire hydrant details, water/fire line details showing meter connection, a water map showing fire hydrant locations used to conduct fire flow test(s) reference to the site, fire lane signage and painting details, fire department connection (FDC) pit details, and any other information related to fire. The fire reviewer should develop a systematic approach to performing a review. A checklist can be created and used as a guide. It is important to note that every site is different and any checklist, no matter how extensive cannot replace a comprehensive review. Section "Site Plan Checklist" in the Appendix provides a general checklist of common items to look for. As a site gets more complicated, such as with podium/pedestal-type structures, where one or more buildings are on top of each other, the review may also encompass different codes, and the checklist may need to be expanded. For example, per NFPA 14, a high-rise building requires two FDCs (also referred to as Siamese connection) for each standpipe zone, which must be remotely located from each other. All FDC locations must be shown on the plans with the serving fire hydrants (more about FDCs in Chapter 6).

According to the t-squared model, a fire doubles in size exponentially; therefore every minute counts. Firefighters rely on a rapid/organized response therefore any delay adds to response time and causes a domino effect in operations. The sooner the units can arrive, the faster they can survey the building and conditions, locate and suppress the fire, and remove the occupants and themselves from harm. At the site plan review stage, the job of a fire plan examiner is to account for obstacles that may hinder this process. During fire incidents, high stress, along with weather and unfavorable site conditions (construction) and unfamiliarity with the area (new volunteers and those on call from other districts), can place the lives of fire personnel in danger (OSHA, 2006). "The number of firefighters dying in the line of duty has remained above 100 annually for all but a few years since NFPA began compiling statistics in the late 1970" (Gorbett and Pharr, 2011, p. 2). An in-depth review of the site plan may not be able to eliminate these deaths, but it can reduce their number, which means lives can be saved. Evaluating for fire features will improve firefighters' chances of success.

BUILDING ADDRESS

Address can play a key role. Determine whether there is an address or multiple addresses if there are multiple buildings or tenant spaces within a building. The address must be clearly stated on the plan. Remember that every emergency unit responds to an address, which is typically composed of a building number and a street name. If the building has many addresses, then they all must be indicated on the site plan.

It is possible to have one building with multiple street addresses, for example, a retail plaza with different tenants. The reviewer must make sure the address makes sense. It may be best to have the address (both building number and street name) on both sides of a building, if the building faces another street. You may need to touch base with other agencies that assign addresses and coordinate with them. There may be times when the address is not assigned at the site plan stage. The International Fire Code (IFC) gives power to the fire official to require/assign addresses. In any case, it is important to reevaluate the site plans if the address changes. Most likely it will be based on the street facing the main entrance. This is where the attending unit initially arrives. As the driver of the fire truck is driving toward the building, he/she should be able to see the building number from the street on which the building is addressed. The FDC should be located on the street front side of the building. When there are multiple sprinkler/standpipe systems in a building, or one of the buildings is partially sprinklered, signage at the FDC needs to be provided to indicate which system it serves ("AUTO SPK," "Standpipes") and which area(s) of the building the FDC is associated with. This is further stated in NFPA 13 and 14 standards. As mentioned above, a retail building can have multiple tenants who may have different addresses, and some spaces may be required to have sprinkler systems while others may not. More information on the FDC is presented in Chapter 6. It may be prudent to have the addresses indicated on the back of the buildings so that responders can make their way from the rear if necessary. Some strip shopping centers in plazas have multiple entry points. Keep in mind that the overall goal is to have responders quickly locate the building without a lengthy approach (ICC, 2011).

In the layout shown in Figure 1.2, townhouses are addressed off the internal private streets on the garage side. The entrances to these homes face the main public

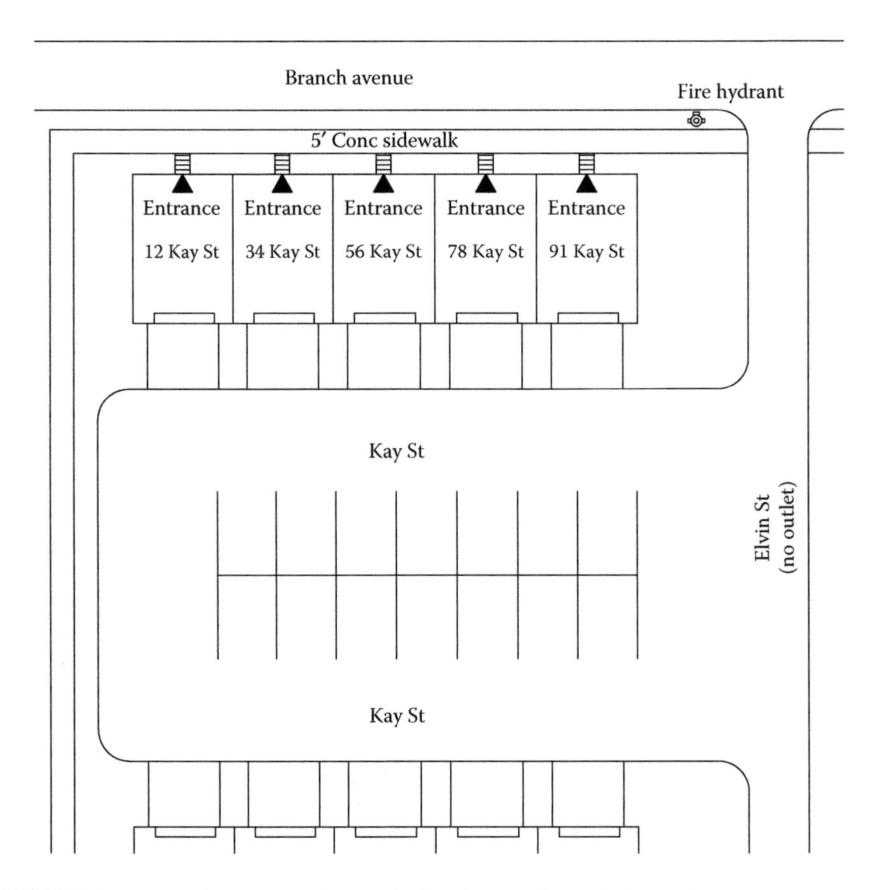

FIGURE 1.2 Townhomes not addressed off main road (Branch Avenue).

street, Branch Avenue. This presents a challenging scenario for the emergency units responding to the address, as they will think the houses are addressed off Branch Avenue. When they arrive on the private road (Kay St), they will find the garage door closed with no access to the house. Here, two options are possible. Either have the town houses addressed off the main street (Branch Avenue), or put a sign above the front door of each house indicating the private street address, that is, 56 Kay St at the door of the town house on the Branch Avenue side to let the arriving units see the address from Branch Avenue. The optimal time to resolve addressing issues is at design. Addressing issues must be resolved for site plan approval.

Figure 1.3 shows an example of pedestal/podium-style buildings. The structure consists of three buildings: the garage, the high-rise, and the low-rise. Having one address for this entire structure can be confusing, resulting in problems such as which building to respond to, how the buildings should be annunciated, the number of FDCs associated with the correct building and at the right location, the water line shut off, etc. At a minimum, the high-rise and low-rise must have separate addresses, with the garage either independent or associated with one of the two buildings.

Parking garages may not have their own address. They are usually associated with another building. Such an example is a parking garage attached to a hotel or a residential or office building.

At times, the main entrance of a building may not be on the address side. The main entrance must be identified on the site plan. The main entrance for buildings may contain the annunciator panel, the fire alarm panel, and the building graphic layout, and can provide direct access to a fire command center, the elevators, etc. During review, one must locate the entrance and determine whether or not suitable access is provided. See Chapter 7 for more information regarding fire department access.

The IFC provides specific information for the identification of premises in terms of letter and number sizing and width, illumination, and coloring. If the site is large or complex, it may be necessary to provide arrows to guide the emergency unit. In some cases, a mini map with the street name and building, showing the location

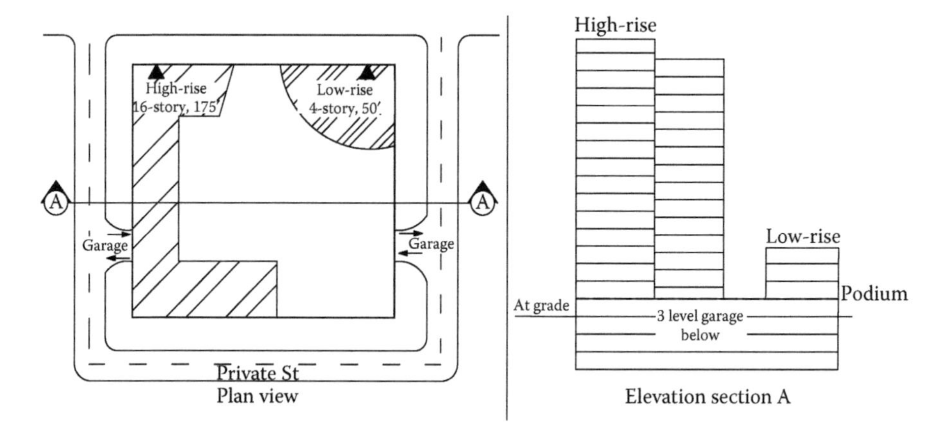

FIGURE 1.3 Multiple buildings on a podium/pedestal.

of fire hydrants, Siamese connections, main entrance, sprinkler valve room, exits, etc., can be put on the face of the building. These elements are beyond fire site plan review, but should be kept in mind. The location of rapid entry boxes (such as a Knox box) containing keys to the premises should be called out on the site. A Knox box is a Underwriters Laboratories (UL) listed box that contains keys to the building, gates, sprinkler room, elevators, etc. that are used by the fire code official to gain entry to the premises when needed.

All roads providing access to the property need to be located on site plan. These can be public or private. Some of these roads provide fire truck access to the building(s), with serving fire hydrants along them and adequate width for equipment setup and turnaround, while others are used for utility or private traffic. Not every road or passageway on a site can be used for fire access. The fire access roads have specific requirements and, if necessary, should be distinguished on the site plan.

MEASURING SCALES

The reviewer needs to be aware of the scale used in the drawings. A graphic representation of the scale used in design should be present on all sheets, with the scale called out in the border (Figure 1.4).

Most site plans are drawn to engineering scales, also known as civil scales. In the United States, an engineering scale is one that represents a unit of measurement per foot. These scales are typically 1:10, 1:20, 1:25, 1:30, 1:40, 1:50, or 1:60, which represent 1 inch measured equals 10 feet, etc. Metric scales tend to be 1:100, 1:200, 1:400, etc., each unit representing a meter length. In order to accurately review site plans, the reviewer must be familiar with these scales (Figure 1.5).

What if one does not have a particular scale? There may be times when you do not have a certain scale available. For example, the scale does not have 1:25. One needs to improvise in this case. Use the 1:50 scale and divide the measurement by two to get a measurement for 1:25. Water and fire line profiles are typically done on

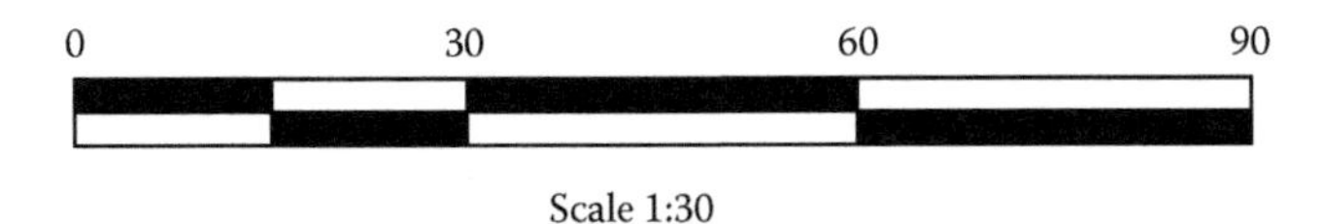

FIGURE 1.4 Representation of a graphic scale found on plans.

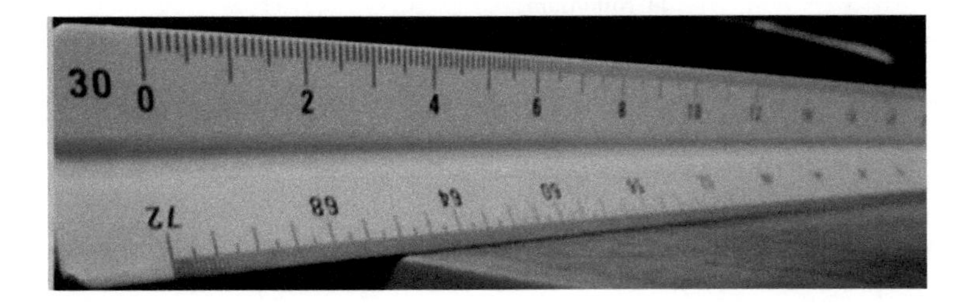

FIGURE 1.5 Engineering scale.

a different scale 1:5 (e.g., vertical 1:5, horizontal 1:20), in which case a scale of 1:50 can be used. Divide all the measured values by a factor of 10 to get 1:5.

PROFILE

A profile is a section view that is used to provide additional detail, typically a cross section of utilities or roads. For a water or fire line, it shows pipe elevation in the ground, the crossing of other utilities, bends, taps, valves, etc. along station markers. Figure 1.6 shows an example of a profile. Chapter 4 will discuss fire line profiles in detail.

EXISTING CONDITION/DEMOLITION

After the coversheet, the reviewer should turn to the existing and/or demolition sheet. Sometimes these sheets can be one in the same, but sometimes they are separate. These sheets provide a picture of what is on currently on site. The reviewer can grasp the magnitude of changes that will be occurring, whether they are minor (such as a storm water line addition) or significant. The reviewer should check the following items:

- If there are any surrounding buildings, the reviewer should ask whether these buildings are sprinklered. If sprinklered, then it is necessary to locate the FDCs and fire lines. Check on the new site layout to see that they remain and are not obstructed or affected. In accordance with NFPA 24 and 13, fire lines should not be built over unless special considerations, as identified by the standards, are met.
- Mark all existing fire hydrants to determine if any are being demolished or relocated. Check to determine if coverage is reduced because of the new layout. Confirm that existing FDCs still have a serving fire hydrant.
- Verify that fire truck access to existing building(s) is still met—that is, width of access road around the site, no dead ends created or grades affected, etc.

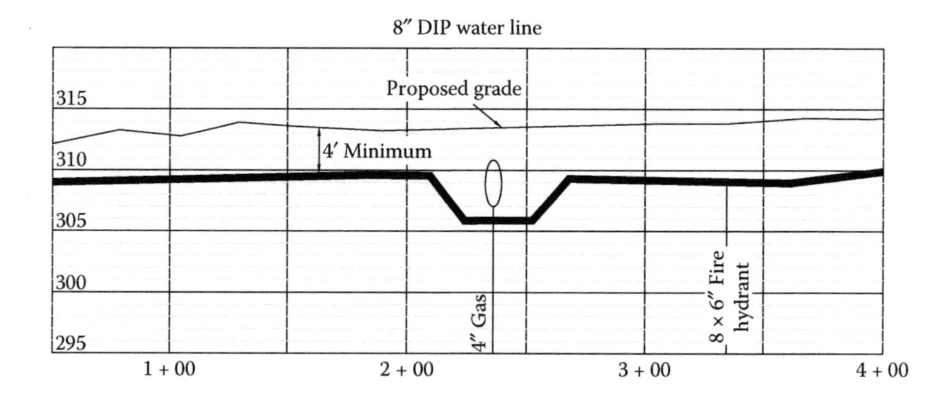

FIGURE 1.6 Water main profile.

- Note the property lines and verify that fire separation distance requirements are met for the new buildings. If the distances are not met, then the exterior fire ratings of the new buildings must be increased.
- Check for underground fuel tanks and pipe lines.

REVISIONS

There are times when a revision to an approved site occurs. A narrative should be provided explaining the changes. The changes should be red-lined and shaded so that the reviewer can easily locate them and not have to reevaluate the entire plan.

SITE RENOVATION

Any modifications to a site should be evaluated by a fire plan examiner to ensure access and ensure that safety is not compromised. While it may seem minor to the lay person, it can have a detrimental effect on fire rescue. One example of this is adding an emergency generator with a diesel fuel tank. Many facilities today are being renovated to provide generators on site to serve as backup power in case of outage, especially in large data centers. The location of the fuel tank(s) needs to be evaluated to ensure that a proper distance is achieved from the building and the property line. These distance requirements can be found in the NFPA 30 *Standard for Flammable and Combustible Liquid*. Furthermore, the tank must not be located over fire lines or in the path of fire access, or near FDC connections or fire hydrants. There have been instances where generators have unintentionally been placed directly on a fire access path (e.g., on grass pavers), and this has reduced fire department access, making the building non-code compliant. The reviewer must also watch out for fences impacting fire hydrant location and hose lay paths (see Chapter 6). He or she must also watch out for additions/expansion to buildings, which must not be over fire lines, and their coverage distance to serving fire hydrants (Chapters 4 and 5), and fire flow must be reassessed for expansions/additions and exposure changes (Chapter 3).

LOCAL FIRE ALARM NOTIFICATION

Buildings with an automatic fire sprinkler system are required to have exterior alarm bells that activate when the system trips. The site plan needs to show the bell located at the exterior of the building, preferably near the entrance, so that anyone approaching the building can stay out, and bystanders can call 911 to notify of an emergency (Figure 1.7).

RECORD KEEPING

Once the review is complete, and if the plans are approved, the reviewer must retain one set of plans as a record. The plan sets are stamped "approved" or "approved as noted." The fire plan examiner should initial and date all the sheets he/she has reviewed that are of fire importance, that is, cover sheet, fire lane sheet, fire line

FIGURE 1.7 Exterior bell on wall. Also shown are control valves with tamper switches.

profiles, grading, fire lanes, etc. The date stamp helps differentiate any revisions or updates of approvals. As site plans can contain many sheets, only copies of the sheets reviewed and signed should be kept to save space (e.g., there is no need to keep the erosion and sediment control or storm water sheets). Along with the site plans, any communications—such as copies of rejection letters, letters addressing comments from the engineer, code modification, or special exception requests—must be kept with the approved site plan. It is best practice to keep record of site plans as long as possible. This will help with any renovations that occur on site. Follow your jurisdiction polices for the length of time necessary for record keeping. Signed copies of plans can be scanned and kept electronically. The reviewer can request the engineering firm to send an electronic copy to retain for the records.

REFERENCES

Gorbett, G. E., and Pharr, J. L. 2011. *Fire Dynamics.* Upper Saddle River, NJ: Prentice Hall.

International Code Council (ICC). 2011. *2012 International Building Code and Commentary.* Country Club Hills, IL: International Code.

NFPA. 2008. *Fire Protection Handbook.* 20th edn. National Fire Protection Association, Quincy, MA: National Fire Protection Association, Inc.

Occupational Safety and Health Administration (OSHA). 2006. *Fire Service Features of Buildings and Fire Protection Systems.* U.S. Department of Labor: OSHA 3256-07N. Retrieved from https://www.osha.gov/Publications/fire_features3256.pdf

2 Grading

You may have heard the term *at grade* with respect to building floor elevations, but what is the importance of "grade" or "grading" with respect to fire site plan review? Almost every site plan should contain a grading plan. Grade is measured as a percentage which represents rise over run. A grade plan shows contour lines that reference changes in ground elevation. *NFPA 1901: Standard for Automotive Fire Apparatus* defines "grade" as "a measurement of the angle used in road design and expressed as a percentage of elevation change over distance" (NFPA, 2008, p. 13). This is measured on the plan from one contour line to another.

$$\text{Grade} = \frac{\text{rise}}{\text{run}} \times 100\%$$

From point A to point B, the surface rises 2 units over a distance of 20 units. Thus, $2/20 = 0.1 \times 100\% = 10\%$ grade. From point B to point C, the ratio is $2/50 \times 100\% = 4\%$ grade. See Figures 2.1 and 2.2.

Think about going up a steep grade. Not only will it tire you out quickly, but it will also not allow a proper equipment setup. A gradual slope allows for a rapid equipment setup, thus allowing the crew to respond faster. The reviewer must note any steep locations.

Grading affects three critical aspects of fire operation activities. The first aspect to consider is pooling of water as it relates to discharge locations at fire hydrants, pump test header locations, or test and drain locations. Inadequate grading at these locations allows water to collect and creates a hazard where emergency personnel may be working. Therefore, a minimum grade should be established to allow water to run off from the appropriate location, and also not to allow the water to freeze (in localities subjected to frost).

The second aspect is to verify the locations at fire hydrants and FDCs to make sure they are accessible and not up/down a steep hill. The location of any retaining walls should also be checked. There have been many instances where fire hydrants, FDCs, and even aerial ladder truck setup areas have been obstructed by retaining walls.

The third aspect to consider is the fire apparatus access road. Fire trucks need accessible roads to drive on to get to the building. The IFC requires road access for fire apparatus to within 150 feet of all portions of the exterior wall. This access is very important as it allows firefighters to get to most locations on the ground floor, gain access to the rear and side exits, break windows to vent the fire, and set up ground or aerial ladders to get to the floor above (discussed in greater detail in Chapters 7 and 8). The right grade allows for fire trucks to traverse smoothly to the site, for aerial ladder trucks to set up outriggers, for firefighters to set up ground ladders at the appropriate angle to the side of the building, and for equipment setup, such as ground monitor nozzles (see Figures 2.3 and 2.4).

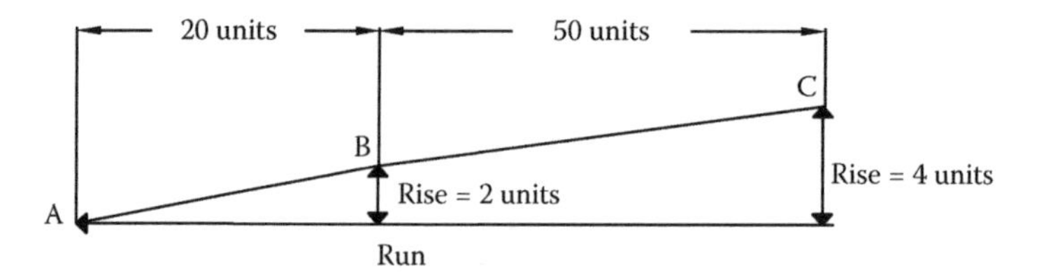

FIGURE 2.1 Section view showing change in grade.

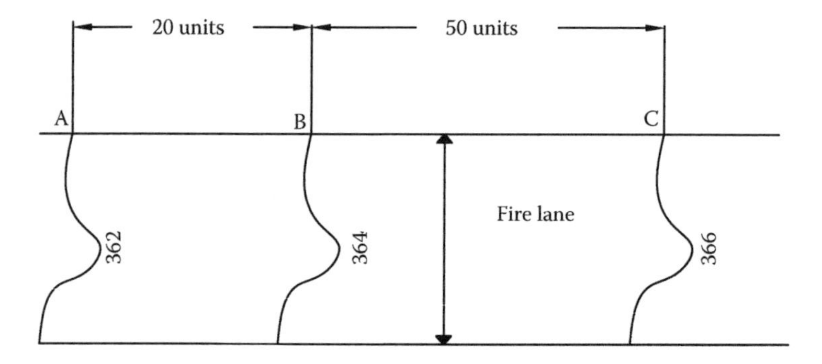

FIGURE 2.2 Plan view showing change in grade.

FIGURE 2.3 Ladder set up against a building wall. (Courtesy of Fairfax County. With permission.)

FIGURE 2.4 Monitor nozzle setup on grade. (Courtesy of Fairfax County. With permission.)

There are several different recommendations on grading for fire access. *NFPA 1142 Standard on Water Supplies for Suburban and Rural Fire Fighting 2007* contains guidelines for acceptable roads leading to water supplies in rural areas (NFPA 2006). A maximum grade of 8% is allowed, whereas NFPA 1 Fire Code states that a maximum grade of 5% is appropriate for fire department vehicles, and IFC recommends a maximum grade of 10%. Fire truck performance is based on a maximum grade of 6%, in line with NFPA 1901. Vehicles can be designed and built for steeper grades upon special request to manufacturer. To determine the parameters for your jurisdiction, check with your operations department on the trucks you have, get the necessary information from the manufacturer, and take into consideration the maximum slope firefighters would need to set up in all weather conditions.

The reviewer should also be aware that any sudden changes in grade can damage a fire truck. Two terms come into play. The first is the angle of departure, which, according to NFPA 1901, is "the smallest angle made between the road surface and a line drawn from the rear point of ground contact of the rear tire to any projection of the apparatus behind the rear axle." The second is the angle of approach, which is

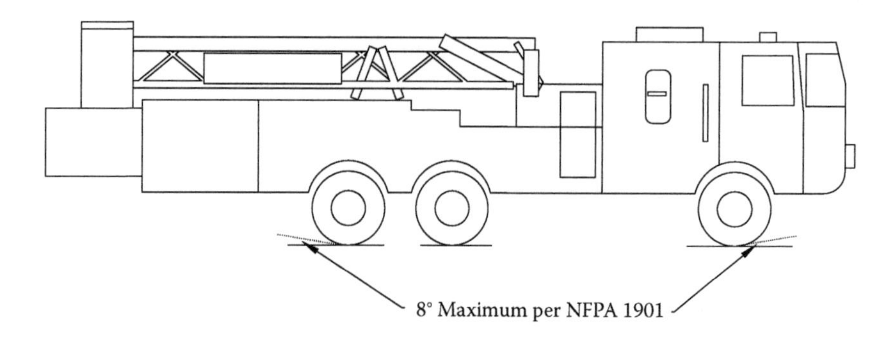

FIGURE 2.5 Image of an aerial platfrom.

"the smallest angle made between the road surface and a line drawn from the front point of ground contact of the front tire to any projection of the apparatus in front of the front axle" (NFPA, 2008). Both of these angles should be a maximum of 8° according to NFPA 1901 (Figure 2.5).

So what is so important about these angles? If the truck has to go up or down a steep hill where this angle is exceeded, then the apparatus can be significantly damaged. The reviewer must evaluate the grading plan, look for areas on the fire truck access road where the contour lines are closer together (which means a rapid change in grade), and request additional information if necessary to confirm that the truck can traverse the access road safely (Figure 2.6).

PUBLIC VERSUS PRIVATE STREETS

If you have not already done so, you will soon be thinking: do these grade requirements apply to public (Department of Transportation) governed streets? According to the IFC, a fire apparatus access road is defined as

> A road that provides fire apparatus access from a fire station to a facility, building or portion thereof. This is a general term inclusive of all other terms such as fire lane, public street, private street, parking lot lane and access roadway. (ICC, 2014)

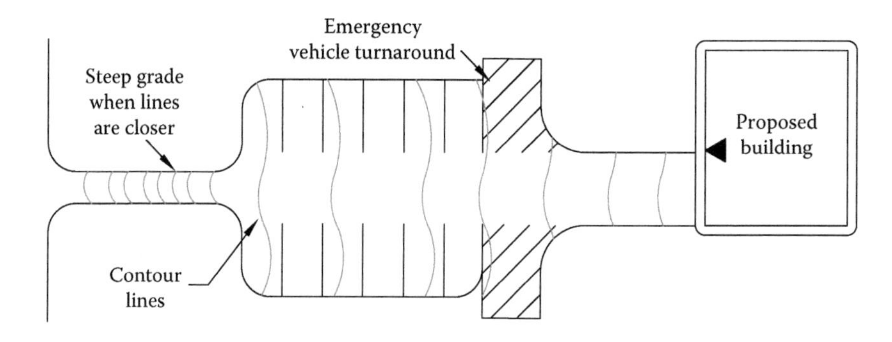

FIGURE 2.6 Example of a site with different grades.

Public roads, state highways, and bridges are part of fire department access, but the IFC grade requirements may not be enforceable on public streets as they fall under federal and state regulations. Typically, these roads are designed for fire apparatus movement, although there may be instances where the grading exceeds the recommendations. See which of the codes or standards your jurisdiction has in place and which of those are higher up in the hierarchy. It is a good idea to have contacts in the federal agencies and state road agencies that you can go to for information.

REFERENCES

International Code Council (ICC). 2014. *International Fire Code 2012*. Country Club Hills, IL: International Code Council, Inc.

NFPA. 2006. *NFPA 1142 Standard on Water Supplies for Suburban and Rural Fire Fighting 2007.* Quincy, MA: National Fire Protection Association, Inc.

NFPA. 2008. *NFPA 1901: Standard for Automotive Fire Apparatus 2009.* Quincy, MA: National Fire Protection Association, Inc.

3 Fire Flow

Fire flow is the water required in gallons per minute (gpm) (or liters per minute) to put out a fire. When the fire pumper truck connects to a water supply and the firefighters are using the hoses of the pumper, fire flow is being utilized. Water has a high heat capacity and is able to absorb a lot of heat. The bigger the fire, the more energy is released, and therefore the more water needed to suppress the heat and extinguish the fire. Fire flow is not to be confused with hose stream requirements for a fire sprinkler system interior to the building. Hose stream is based on water needed for manual firefighting while the sprinklers are in operation inside the building. Fire flow encompasses the water demand for the entire building on fire used solely for exterior firefighting. There are many variables involved with fire flow analysis and several methods available for determining the fire flow. Some of the most popular ones used to determine the flow are the Insurance Service Office (ISO) method, the Iowa State University (ISU) method, the Illinois Institute of Technology Research Institute method, and the *NFPA 291: Recommended Practice for Fire Flow Testing and Marking of Hydrants* method, and the appendix of the IFC also provides a chart of fire flow needed on the basis of the area of the building and the construction type. For rural areas, fire flow is addressed by NFPA 1142. Fire flows were originally developed for nonsprinklered buildings, but due to instances where the sprinkler system was inoperable/ineffective, fully sprinklered buildings get a reduction. For example, the IFC appendix introduced a 50% reduction in fire flow for one and two single-family dwellings, but a 75% reduction for others that are fully sprinklered. It would be unreasonable to ask for fire flows for existing buildings as suitable water distribution systems might not have been present in the past. In fact, existing building codes may prohibit not only this but also requests to increase fire access to site or additional hydrants. The methods discussed in this chapter apply to new buildings or modified sites (e.g., change in property lines increasing exposure distance) or for buildings that are being expanded with additions.

ISU METHOD

This is the simplest of all the methods. It is based on water converted to steam to replace oxygen. It takes into account the entire volume of confined spaces in the building.

$$\text{Required fire flow in gpm (gallons per minute)} = \frac{\text{volume (in feet}^3\text{)}}{100}$$

ILLINOIS INSTITUTE OF TECHNOLOGY RESEARCH INSTITUTE METHOD

This method is a regression analysis composed of over 134 fire surveys conducted in Chicago (NFPA, 2003).

For residential occupancies:

$$\text{Required fire flow (gpm)} = 9 \times 10^{-5} \times \text{A}^2 + 50 \times 10^{-2} \times \text{A}$$

For nonresidential occupancies:

$$\text{Required fire flow (gpm)} = -1.3 \times 10^{-5} \times \text{A}^2 + 42 \times 10^{-2} \times \text{A}$$

"A" represents the gross floor area (sum of all the floors) in feet^2 of the building.

ISO METHOD

This is probably the most popular method and involves taking into account many factors, such as building construction type, use group, fire separation distances, communication space, gross floor area of the building, and exposure to other buildings. The calculations use tables, several formulas with many intermediary steps making it complex. See the NFPA *Fire Protection Handbook* for more details. For single-family dwelling units that are not more than two stories in height, a table is provided for acceptable fire flows needed.

FAIRFAX COUNTY METHOD

It is possible to adopt any of the recognized methods above for use at fire site plan review. Fairfax County adopted the ISO method with some modifications. The formula was compressed to reduce the time needed to perform the calculations. Furthermore, fire flow is required for every building or home, even those that are fully sprinklered throughout. A fully sprinklered building is given a 50% reduction in the calculated fire flow. One can obviously see the benefits of a sprinkler system, as it provides automatic fire control so why is a reduction applied? Instead why not completely eliminate the needed fire flow requirements for a building with an automatic sprinkler system as there will be no need for manual firefighting? Answer: as indicated in the Analysis section, inadequate maintenance and under design of the fire suppression system results in failure of the system to control and suppress the fire. Furthermore, recall the apartment building fire at apartments in Rockville, Maryland, described in Introduction section. Having a fire flow for manual firefighting in the event of a sprinkler system failure is the last chance available to save the building and occupants. Additionally, fire sprinkler protection does not suppress a fire from the exterior.

The formula used is as follows:

$$\text{F} = 18 \times \text{C} \times \text{A}^{1/2} \times \text{OR} \times \text{ED} \times \text{sprinkler reduction}$$

where

F = Fire flow in gpm
C = Coefficient factor based on construction type
A = Total area of all the floors (gross floor area) not including basement
OR = Occupancy reduction
ED = Exposure distance
Sprinkler reduction = 50% or 0.5 and it is only for buildings that are fully sprinklered per the building code

Type of Construction (per IBC 2009)	Coefficient—"C" Value
Wood frame (VA, VB)	1.5
Ordinary (IIIA, IIIB)	1.0
Heavy timber (IV)	0.9
Noncombustible (IIA, IIB)	0.8
Fire resistive (IA, IB)	0.6

Occupancy Reduction—"OR" Value

Owing to the nature of certain occupancies, further reductions can be applied. This is mainly because of the presence of low combustibles, compartmentation, or simply because a fire alarm and detection system is required by code to be present. A fire alarm system and detection these systems allows for early detection of a growing fire and it means that the fire department can get to the site sooner and control the fire with less water as it will not be as large. Further, the alarm will alert the occupants to leave the building. Below is the chart taken from the Fairfax County *Public Facilities Manual.*

Type of Occupancy	%	Type of Occupancy	%
Asylums	15	Prisons	10
Churches	15	Public buildings	10
Clubs	10	Rooming houses	10
Dormitories	25	Schools	15
Hotels	20	Parking garages (stand-alone, not under buildings)	25
Hospitals	10		
Nursing homes	15		
Office buildings	10		

The critical thing to remember is that this is a reduction, so the fire flow will be reduced by this percentage. For example, if the calculated fire flow is 2,000 gpm before "OR" for an office building, it can be reduced by 10%.

$$2{,}000 \text{ gpm} \times 10\% \text{ (or } 0.1) = 200 \text{ gpm}$$

$$\text{New required fire flow} = 2{,}000 \text{ gpm} - 200 \text{ gpm} = 1{,}800 \text{ gpm}$$

A faster way of doing this is 2,000 gpm $\times$ (100% $-$ 10%) = 2,000 gpm

$\times$ 90%(or 0.9) = 1,800 gpm

Exposure Distance—"ED" Value

The exposure distance value represents the hazard to adjacent building(s) from the building on fire. It is the separation distance necessary to reduce the effects of fire. Factors such as radiant heat and flying embers landing on adjacent building(s) are considered. While building codes consider fire separation with regard to wall ratings in terms of measurement to the lot line, to an imaginary line between two buildings, or to the center line of the public way, this ED value accounts for fire flow based on the ISO method and is more stringent as it accounts for buildings that are as far as 150 feet of the reference building. Why 150 feet? Recall the Great Chicago fire, where high winds blew embers and burning debris several hundred feet. As you can see, the ED does not factor in the construction type or use group of the building. Since water is not efficient at absorbing radiant heat, a method to prevent exposure relies on running water down the exposed wall, thereby conducting heat away from the surface as it flows (like a water curtain), keeping the surface relatively cool (Shackelford, 2009). Any communication spaces, such as external connecting corridors from one building to another, must have fire walls rated to at least the same duration as the necessary fire flow. If not, the gross floor area used in the formula above must be extended to the adjacent building. If there is a true separation, and if rated fire walls/exterior walls are present on the exposed building surface, then the exposure percentage is 0%. Below is the chart taken from the Fairfax County Public Facilities Manual.

Separation (feet)	Percentage (%)
0–10	25
10.1–30	20
30.1–60	15
60.1–100	10
100.1–150	5

This is a compressed version of the ISO chart. The maximum ED per ISO should not be more than 75% combined for all sides. The distance is measured perpendicular from the wall to the face of the adjacent building. Furthermore, the building code in place may also require the exterior rating of the walls based on the fire separation distance. For one- and two-story family dwellings, both the ISO and IFC set out a similar procedure to obtain the required fire flow for not more than two stories. See the following table:

Exposure Distance (feet)	Fire Flow (gpm)
0–10	2,000–1,500
11–30	1,500–1,000
31–100	1,000–500
Over 100	500

The closer one house is to another, the greater is the fire flow (gpm) required. The ED is measured to the property lines, the boundary limits of each property. It is not that the property lines will catch fire; rather, it is because home owners can put storage sheds, build gazebos, or even expand their houses anywhere on their property up to the property line. To be more accurate, linear approximations can be used to find fire flow in between the ranges. Fire flow for townhouses needs to consider whether or not rated fire walls exist of equal fire flow duration between each unit. If there are fire walls, then one must calculate fire flow for the largest unit only, as they will prevent flame spread.

IFC METHOD

Another available method for determining fire flow is from the IFC appendix. This is not part of the code, but can be adopted by any jurisdiction. The table (used for any building except for one- and two-story family dwellings with an area less than or equal to 3,600 feet^2) is based on ISO guidelines from 1972 and can make for extreme and excessive fire flow requirements, with the exception that it allows for sprinkler reduction of up to 75% (50% reduction for one- and two-story family dwelling units). The table takes into account the largest fire flow area (gross floor area of the building separated by fire walls) and construction type, but no occupancy considerations. Maximum fire flow for fire-resistant noncombustible construction is 6,000 gpm, rising to 8,000 gpm for combustible construction. For buildings of multiple construction types, the fire flow can be based on a combined percentage approach (ICC, 2009). For one- and two-story family dwellings of up to 3,600 feet^2, maximum fire flow is 1,000 gpm for a 1 hour duration with a 50% reduction if the building has an approved automatic sprinkler system.

MINIMUM AND MAXIMUM FIRE FLOWS

The ISO method indicates that, for combustible construction, the maximum fire flow should be limited to 8,000 gpm, while for noncombustible construction, the figure is 6,000 gpm, whatever the area (similar to the IFC method). Sprinkler, occupancy reduction, and surcharges should be applied to these values when applicable. Nevertheless, the fire flow should be never less than 500 gpm or more than 12,000 gpm (NFPA, 2008). The amount of water needed is expressed in gallons per minute (gpm) at 20 psi residual pressure for a duration ranging from 2 to 4 hours. The minimum needed fire flow for any single building is 500 gpm for 2 hours. The maximum needed fire flow is 12,000 gpm for 4 hours (Hickey, 2008, p. 69).

EXAMPLE OF THE FAIRFAX COUNTY METHOD

A high-rise office building of noncombustible fire-resistive construction is twenty-five stories tall, with each floor of 70,000 feet^2, and is fully sprinklered. There are two buildings located on the side, one at a distance of 40 feet and the other at 60 feet (Figure 3.1). What is the required fire flow?

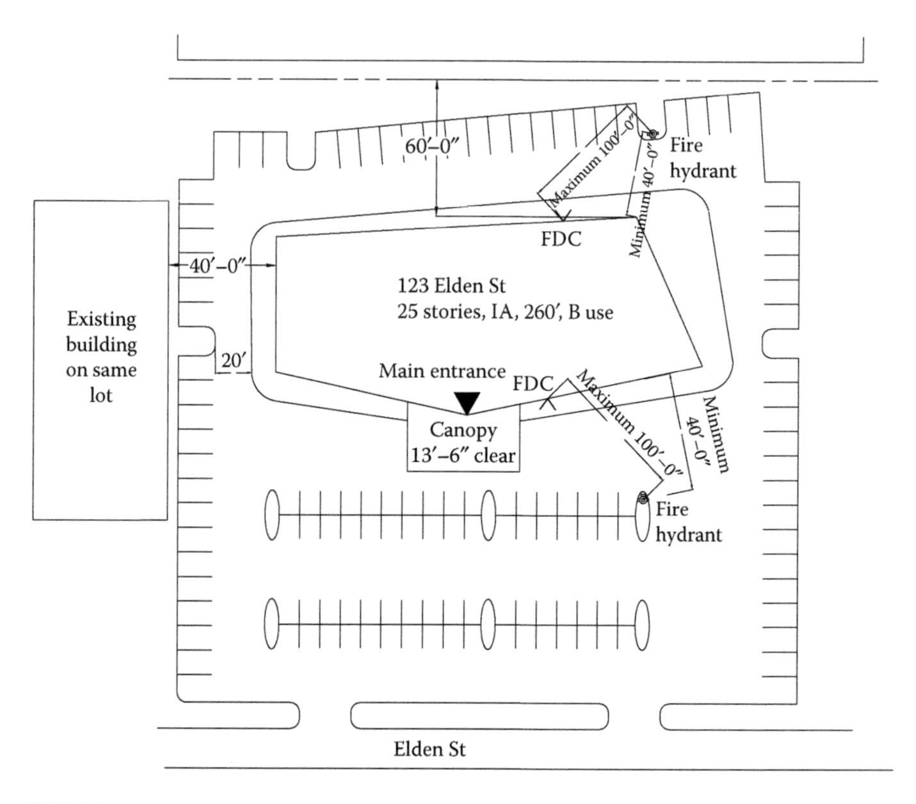

FIGURE 3.1 Proposed high-rise building.

Using the formula:

$$F = 18 \times C \times A^{1/2} \times OR \times ED \times \text{sprinkler reduction } (50\%)$$

$$OR = 100\% - 10\% = 90\% \quad \text{or} \quad 0.9$$

$$ED = 15\% + 15\% = 30\% \quad \text{or} \quad 0.3\text{, therefore an increase of } 1.30$$

$$F = 18 \times 0.6 \times (25 \times 70{,}000)^{1/2} \times 0.9 \times 1.3 \times 0.5$$

$$= 8{,}358 \text{ gpm (in any case, remember to always round up to the nearest whole number)}$$

However, the maximum for any noncombustible building is 6,000 gpm, so 6,000 gpm × 0.5 (for sprinkler reduction) × 0.9 (OR) = 2,700 gpm required.

Both ISO and IFC indicate that in no case shall the fire flow be lower than 500 gpm for any building, to allow for some manual firefighting. Fire flow should be made available for the largest area of the building, or the largest fire area in the building that is separated by fire walls rated for the required fire flow duration. For example, a fire flow duration of 3 hours for one fire area must have fire walls rated for at least 3 hours' duration. If not, the fire flow has to be increased to include the adjacent fire area. There have been instances where engineers/architects argue as to having the fire station right across the street from the building or having the fire tankers supply the water for fire flow or part of the fire flow. Note that the fire protection features in the codes and standards do not take into account where the building is located with reference to the fire station. Furthermore, it is not possible to know which station is going to respond or how much manpower is available or whether the nearby fire trucks will be occupied elsewhere. The water-carrying capacity of tankers can range from 1,500 to 4,000 gallons (Pierce Manufacturing, Inc., 2013). Seeing that, at a minimum, fire flow is required for 2 hours, a gallons-per-minute results in an addition of 400 gpm at best (4,000 gallons/120 minutes). Fire flows must not be adjusted for these factors. An on-site supply of water must be available at all times.

When sized correctly, water for fire flow can be made available through a public water system via fire hydrants or through stand-alone water tanks, reservoirs, etc. The supply must be reliable and maintained; for example, if using a lake, then freezing temperatures need to be considered in areas where temperature fall below 32°F (0°C). In rural areas where public water supply is not readily available, a lake or a large pond located near the facility (provided that it has sufficient water) can be used with a vertical turbine submersible fire pump to draw water and feed the fire hydrants on site. Additionally, water tanks can be used to supply the fire flow for the required duration. The minimum duration is 2 hours for as little as 1,000 gpm, all the way up to 10 hours for 12,000 gpm (NFPA, 2003). Water tanks are designed in accordance with NFPA 22. In the example above, it was shown that 2,700 gpm is required for the high-rise. On the basis of the *Fire Protection Handbook*, the amount of water required will be 486,000 gallons (2,700 gpm for 3 hours). This is a lot of water, and therefore the public water system is most suitable for this fire flow.

For a public water system, the available fire flow can be measured by a fire hydrant flow test conducted in the field. This can be performed by the water authority or anyone the water authority has designated to conduct a hydrant flow test. The fire hydrant test should not be more than one year old. The reason for this is, because in areas that are going through major developments or have gone through major developments, the fire flow can be reduced. Additionally over time corrosion can impact the internal diameter of the pipe. Both the NFPA *Fire Protection Handbook* and the *Fire Protection Engineering Handbook* discuss the long-term effects of a reduction in a variable known as the "C" factor, which is the pipe roughness coefficient. A lower "C" factor of, say, 80 will cause greater friction loss against a "C" factor of 140, which is used for a new pipe. Conducting the test within one year will give the most accurate results. This is mentioned also in NFPA 14. The fire flow available should be noted on the cover sheet, with a map showing the fire hydrants that were used to conduct the flow test, the size of water lines, and street names.

Fire flow must account for daily domestic use. Water must be present during the worst-case conditions, so it is important to perform a flow test during peak usage time, preferably early in the morning when people are getting up, taking showers, etc., or after work/school hours. NFPA 13 *Installation of Sprinkler Systems* 2007 factors domestic water usage when designing sprinkler systems. It states:

> The volume and pressure of a public water supply shall be determined from water flow test data. An adjustment to the water flow test data to account for daily and seasonal fluctuations, possible interruption by flood or ice conditions, large simultaneous industrial use, future demand on the water supply system, or any other condition that could affect the water supply shall be made as appropriate. (NFPA, 2006)

The same principle applies to fire flow. If a fire occurs in a building, then it is, most likely that the people inside are going to be evacuating and not using water; however, a shower could be left on, or perhaps a lawn sprinkler system could be running. The best approach is to have a conservative design that takes into account the daily consumption factors. One technique is to adjust the water for a low hydraulic gradient. A low hydraulic gradient represents the lowest point that the supplying water tank will reach. This accounts for absolutely the worst-case scenario. This information can be obtained from the water authority.

A serious but frowned-upon statement is that "a building should not be built if there is no water available to protect it". The fire safety/protection is just like any other feature of the building. Exits, lighting, mechanical systems, plumbing, etc. are required by code to sustain the building, so fire protection should be treated no differently. Many considerations can be taken to decrease the required fire flow, such as adding a sprinkler system, increasing the construction specification, adding fire walls to give smaller fire areas, or reducing the size of the building. In areas where public water is not available, NFPA 1142 can be utilized. For fire flow testing methods for fire hydrants, consult *NFPA 291: Recommended Practice for Fire Flow Testing and Marking of Hydrants.*

One last item to point out, although it is beyond the scope of the plan review, is that water for fire flow must be available to the development site before combustibles are present. This also applies to access for fire department vehicles. There have been many instances where buildings have burned down during construction due to the unavailability of water and/or site access.

REFERENCES

Hickey, H. 2008. *Water Supply Systems and Evaluation Methods—Volume I: Water Supply System.* U.S. Fire Administration FEMA. http://www.usfa.fema.gov/downloads/pdf/publications/Water_Supply_Systems_Volume_I.pdf

International Code Council (ICC). 2009. *International Fire Code 2009.* Country Club Hills, IL: International Code Council, Inc.

NFPA. 2003. *Fire Protection Handbook.* 19th edn. Quincy, MA: National Fire Protection Association, Inc.

NFPA. 2006. *NFPA 13: Installation of Sprinkler Systems 2007.* Quincy, MA: National Fire Protection Association, Inc.

NFPA. 2008. *Fire Protection Handbook.* 20th edn. Quincy, MA: National Fire Protection Association, Inc.

Pierce Manufacturing, Inc. 2013. *Tankers.* Retrieved October 2, 2014 from http://www.piercemfg.com/en/trucks/commercial/tankers.aspx

Shackelford, R. 2009. *Fire Behavior and Combustion Process.* Clifton Park, NY: Delmar Cengage Learning.

4 Fire Hydrants

Whether they are used to draw fire flow, to overhaul, or for supplying the FDC via the pumper, the purpose of fire hydrants is to provide immediate reliable water on site for firefighting operations. In this chapter, we will focus on the fire hydrants required on site for buildings. There are two types of fire hydrants, dry and wet.

DRY HYDRANTS

According to *NFPA 1142 Standard on Water Supplies for Suburban and Rural Fire Fighting*:

> Dry hydrant is an arrangement of pipe permanently connected to a water source other than a piped, pressurized water supply system that provides a ready means of water supply for fire-fighting purposes and that utilizes the drafting (suction) capability of a fire department pump. (NFPA, 2006a)

The key point to note about a dry hydrant is that water has to be drawn from the supply (lake, pond, reservoir, tank) by a fire pumper through a pipe that is typically located underground. A dry hydrant must have a reliable water supply (Figure 4.1).

Normally, dry fire hydrants are found in rural areas where there is no public water distribution system or there is only a limited water supply. The source of water for these dry fire hydrants can be large bodies of water, ponds, or water storage tanks (NFPA 22); even swimming pools can be used in some instances. Location-specific concerns such as freezing temperatures, supply contamination, and water level draw-down during drought must be addressed at system design.

NFPA 1142 contains excellent information for dry fire hydrants and their water supplies. Where dry hydrants are needed or utilized, this standard should be consulted.

WET HYDRANTS

A wet fire hydrant is maintained under pressure so that when the hydrant valve is opened, the water discharges from the outlet. In areas subjected to freezing, a wet fire hydrant also known as "dry barrel" is used. This is not to be confused with dry hydrants. The dry barrel allows the vertical portion of the fire hydrant to remain without water until the valve is opened. After closing the valve, any water in the hydrant drains out the bottom automatically (Figure 4.2).

A dry fire hydrant is also not to be mistaken with a wet fire hydrant with a dry run pipe attached to it. To sum up, a wet fire hydrant always has an automatic water supply, while a dry hydrant must have water drawn through it. Consider the following case:

An open parking garage two stories above grade is to be built with a high-rise building on top. Fire truck access is required on top of this garage to serve the high-rise building. The garage must support the weight of the fire truck (covered in Chapter 7),

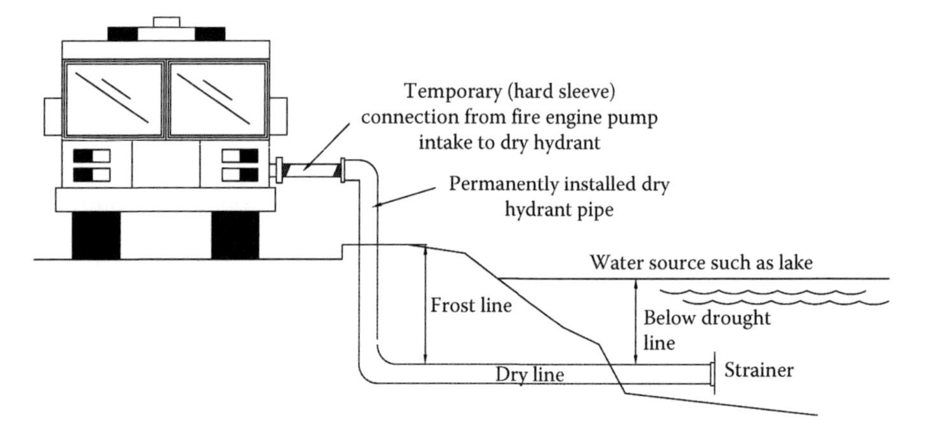

FIGURE 4.1 Fire pumper truck drawing water from lake via a dry hydrant.

and a fire hydrant is needed on the deck to meet the fire hydrant coverage requirements. The pipe supplying the fire hydrant on the deck must run from the street main up to the garage, under the slab, exposed to freezing temperature during winter time.

Problem: If the pipe is filled with water, then, during the winter, it will freeze. Furthermore, building a heated chase would be expensive and not reasonable for this run of pipe.

Solution: The first option is to heat-tape the pipe run exposed in the open garage and provide monitoring for integrity. The reviewer must ensure that the heat tape is listed for fire protection use and the type of pipe (e.g., steel, ductile iron). There are certain restrictions on the type of pipe and size for heat tape. The second option is to have a dry run of pipe from the fire hydrant on the deck running under the slab of the deck (in the open parking garage) with a control valve at the water source in a conditioned space, easily accessible by the fire department; the opening of the valve

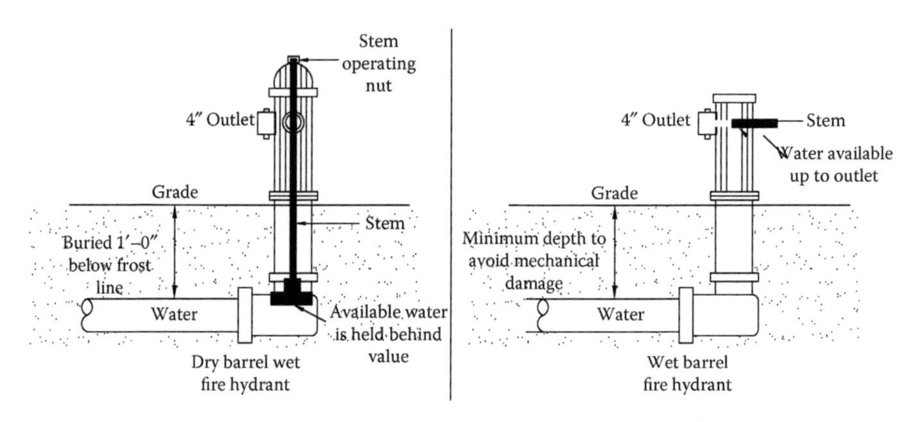

FIGURE 4.2 Section through dry fire hydrant: Both hydrants are wet hydrants but with different barrel arrangements. Dry barrel is used in areas where freezing is of concern. (Adapted from Sturzenbecker, M. J., Adams, B., and Burnside, E. [Eds.]. 2012. *Fire Detection and Suppression Systems*. 4th edn. Stillwater, OK: Fire Protection Publications.)

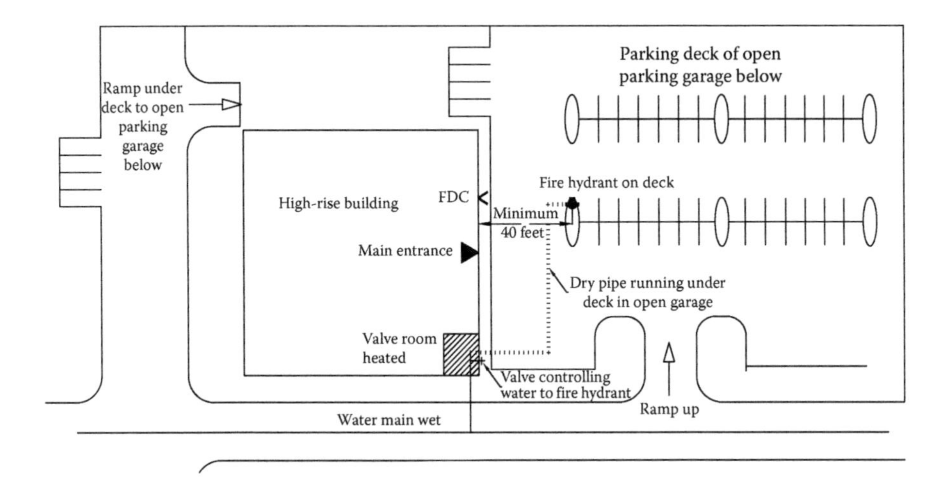

FIGURE 4.3 Fire hydrant example.

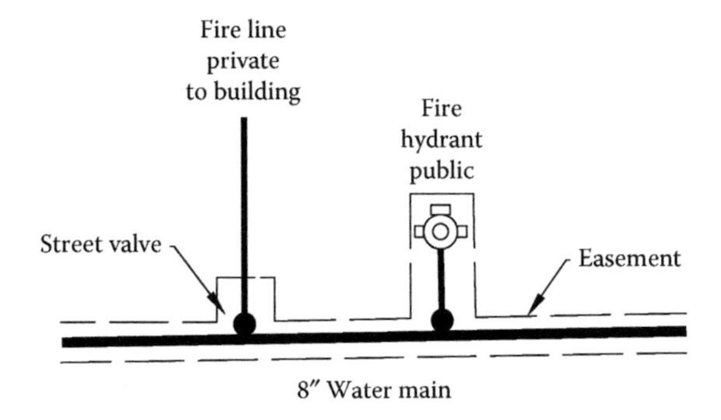

FIGURE 4.4 Easement shown around water pipe owned by water authority.

will allow water to flow to the fire hydrant. A means of draining the pipe in the open garage must be present (Figure 4.3).

Fire hydrants can be public, meaning they are served by a municipal agency or water authority that provides the domestic potable (drinking) water, or they can be privately owned and maintained. Often, private fire hydrants are extensions from the public water system, which means they are beyond the water authority's easement (Figure 4.4).

For information on easements, see Chapter 5. Everything after the street valve must be maintained by the property owner.

FLOW AND PRESSURE AT HYDRANTS

NFPA 24, Standard for the Installation of Private Fire Service Mains and Their Appurtenances provides a set of guidelines for classifying fire hydrants. Flow through fire hydrants is measured in gpm and is based on a minimum residual of 20 psi at the fire hydrant. For public systems supplying potable water, state health departments typically require 20 psi to be maintained throughout the water distribution system. Therefore the lowest pressure at a fire hydrant must be 20 psi, also referred to as Q_{20}, where Q represents the flow and 20 represents the pressure. Fire trucks can perform at pressures lower than these, and even negative pressures (for suction–drafting), but this can damage the public water system and contaminate the potable water.

For public fire hydrants, the size of the pipe serving the fire hydrant is determined by the water authority. A 6-inch or greater branch from the water main is recommended, which is also stated to be the minimum pipe size for private mains supplying fire hydrants per NFPA 24.

Where higher fire flows are required, the supply line may be larger. In the past, smaller pipe sizes were used, but this has been changed by water authorities due to water quality issues. With fire flow under optimal conditions, a 6-inch water main can supply a fire hydrant with up to 1,500 gpm of water, an 8-inch water main can supply 2,500–2,700 gpm, and a 12-inch water main can supply up to 3,500 gpm. All of these values are based on Q_{20}.

The reviewer must request the water authority to provide results for a water flow test for the specific fire hydrant serving the site, or have a flow test conducted in accordance with NFPA 291. On the basis of the results, the flow must be adjusted for a pressure of 20 psi. The general formula to use is

$$Q_{20} = Q_{measured} \times \left[\frac{(Ps - 20)}{(Ps - Pr)} \right]^{0.54}$$

where

Ps = Static pressure in psi
Pr = Residual pressure in psi
$Q_{measured}$ = Flow at the discharge fire hydrant (i.e., flow adjusted based on the pitot tube reading) in gpm

Instead of using the formula, the log graph of N1.85 can also be used to interpolate for Q_{20} (Figure 4.5).

If the required fire flow calculated (Chapter 3) is greater than the available fire flow (flow test from fire hydrant), several actions can be taken. The water main system can be laid out in a loop design, which provides water from two directions to the fire hydrant. The size of the pipes serving the fire hydrant can be increased. Additional hydrants can be tapped from the same street main or from another water pressure zone. For example, an additional fire hydrant can be branched off from the main, but only if the main is large enough to provide additional flow. Here is an example of adding an additional fire hydrant from the same street main (Figure 4.6).

A 6-inch water main in the street will supply up to 1,500 gpm at 20 psi unless indicated otherwise by the flow test. Pulling additional branches of hydrants from a 6-inch water main is of no benefit because there is no more flow available. However, if the hydrants are branched from a larger water main, say, 24-inch, then they can be summed up to a total not exceeding what can be supplied by the 24-inch water main (Figure 4.7).

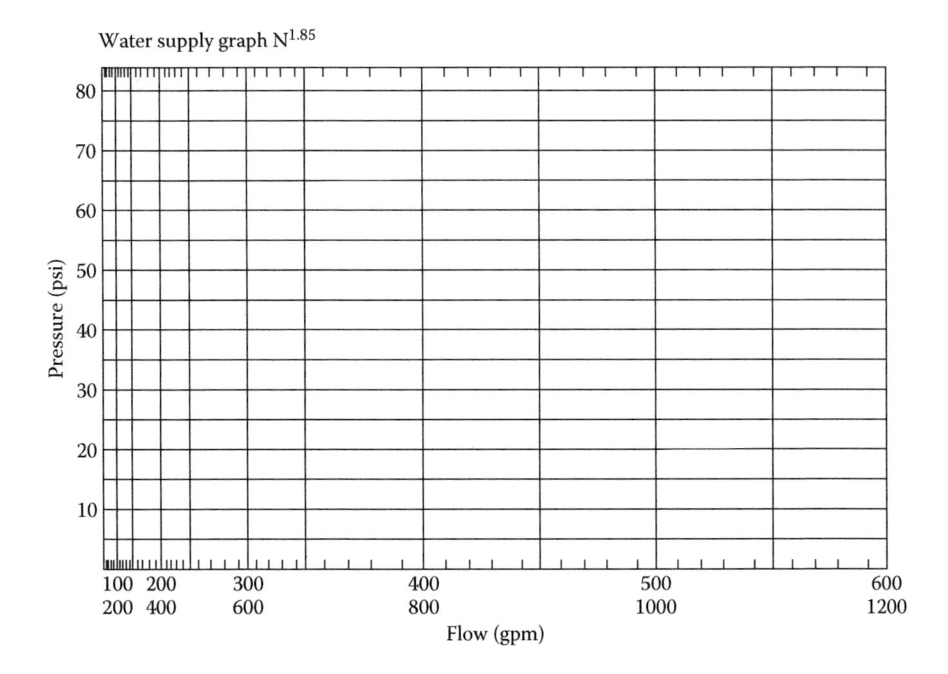

FIGURE 4.5 Example of graph used to plot water supply information.

This is the law of conservation of mass. Mass in equals mass out, which is the same as flow in = flow out or $Q_{in} = Q_{out}$. If the required fire flow is not met, water tanks can be provided on site. The fire pumper suctioning water from the fire hydrant cannot create more flow (i.e., water); pumps/pumpers increase the pressure only. This can be seen on the log graph by plotting the supply with the combined pump curve similar to that of on-site fire pumps for automatic fire sprinklers.

Fire hydrants at the end of the water system (referred to as dead-end fire hydrants) are not recommended due to water quality issues. One way to deal with long stretches of dead end is to upsize the pipe serving the fire hydrant. Another way is to add a check valve in the line, which will allow water to flow in one direction only, toward the hydrant. Unless the hydrant is opened, the water stagnates between the last takeoff and the hydrant. The check valve prevents stagnant water from reentering the potable water supply. Fire flow must be verified to ensure it is adequate for the end hydrant (Figure 4.8).

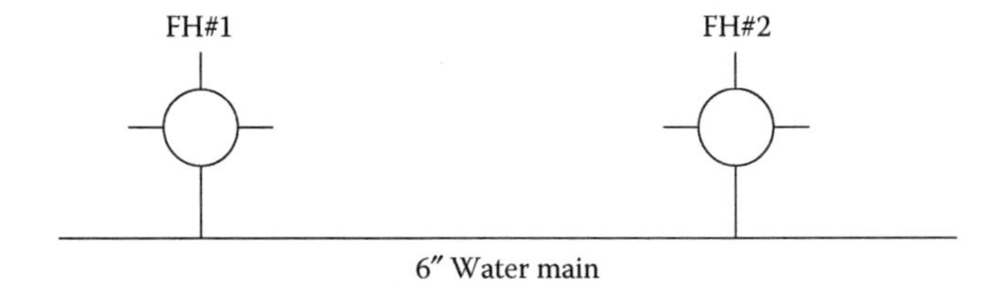

FIGURE 4.6 6-inch water main supplying fire hydrant.

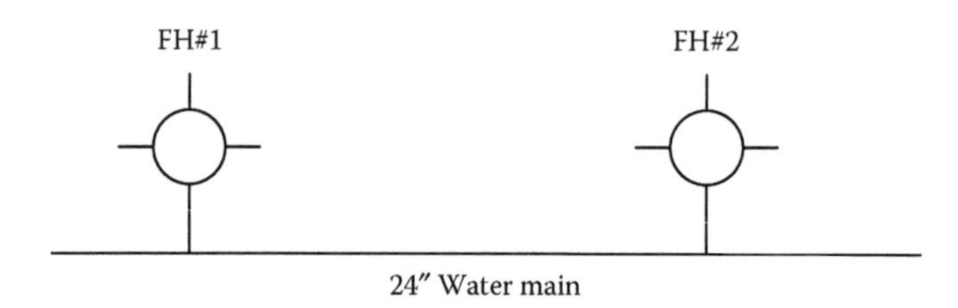

FIGURE 4.7 24-inch water main supplying fire hydrant.

As mentioned earlier, at a minimum, all new fire hydrants should be supplied by a 6-inch main. The largest outlet on the fire hydrant is typically 4.5 inch, and should face the street, so that easy connection can be made to the pumper. The 2.5-inch outlets to the sides supply 250 gpm each, with the 4.5-inch supplying 1,000 gpm (NFPA, 2003, pp. 10–48). See Figures A.4 and A.5 in the Appendix for examples of fire hydrants with recommended reference to curb and pavement location. Fire pumpers carry both hard and soft suction sleeves (short hose) to connect to the fire hydrants. Hard sleeves are used on dry hydrants when drafting water, while soft ones are used on wet hydrants. You should determine the length of sleeves carried by your fire department to give you an idea of how far fire hydrants can be located from the road. It is good practice to locate fire hydrants a maximum of 30 inches from the curb, or approximately 6 feet from traffic lanes when no curb is present. The bottom of the safety flange should be at least 2.5 inches above the grade where it is set. Fire hydrants at intersections should be installed 5 feet from the point of curvature of the street intersection to avoid damage from vehicles. (Figures A.4 and A.5 in the Appendix). The details of fire hydrant dimensions from the curb must be provided on a detail page (Figure 4.9).

Although it does not affect the function of the fire hydrants, the reviewer should verify that the hydrants are not placed in sidewalks, as they do not look appealing, and most likely another agency will prohibit it.

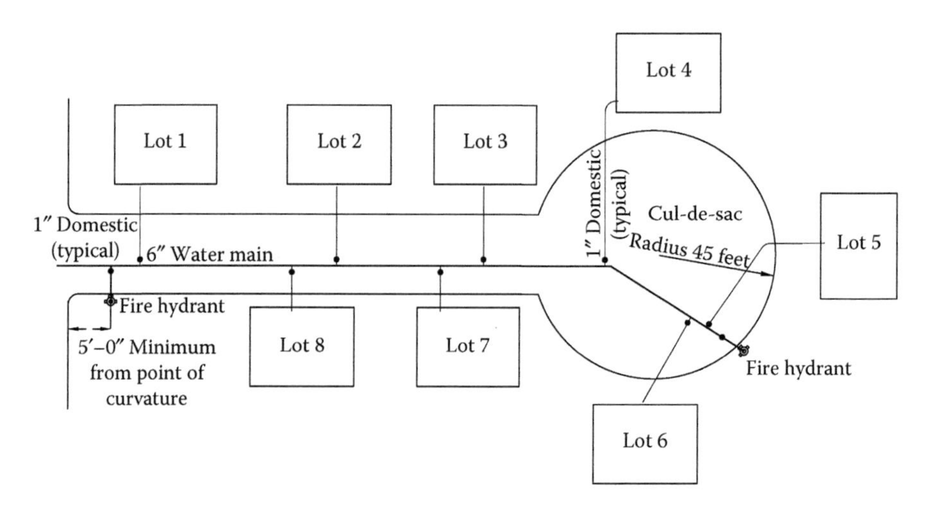

FIGURE 4.8 Location of fire hydrant with respect to access.

FIGURE 4.9 Fire hydrant in sidewalk.

CLEAR SPACE AROUND FIRE HYDRANTS

A 36-inch clear space should be maintained around the circumference of fire hydrants. This is to make sure they are visible and to allow working space for both maintenance and firefighters. Bushes, plants, trees, fences, poles, and snow around the fire hydrant should be kept below line of sight. There should be a minimum grade established so that water does not accumulate after hydrant operation to create hazard conditions. Check on plans to make sure the fire hydrant is not surrounded by car parking that obstructs it (Figure 4.10).

Furthermore, see if the fire hydrant location makes sense. Questions to think about are: is the fire hydrant easily accessible? Would there be any kinks in the hose connection from the pumper to the hydrants? Fire hydrants need to be protected from damage where there is no curb or where other damage can occur; in this case, bollards meeting the necessary code can be used (Figures 4.11 and 4.12).

Bollards can be designed based on vehicle impact protection, as indicated in the IFC. Details like these should be included in the plan set, preferably on the detail sheet.

In Figure 4.13, the parked van does not allow the pumper to achieve a proper connection to the hydrant, making it useless. This is why parking is not allowed near a fire hydrant. The IFC requires that no obstruction prohibit access for the fire

FIGURE 4.10 Car parked near fire hydrant. (Courtesy of Fairfax County. With permission.)

department. Many jurisdictions require no parking within 15 feet of a fire hydrant, which is marked by signage and painting.

DISTANCE BETWEEN FIRE HYDRANTS

What dictates the locations of fire hydrants? How far apart should the fire hydrant be from each other? How many should serve a building? In the past, many fire hydrant

FIGURE 4.11 Barriers around fire hydrant. (Courtesy of Fairfax County. With permission.)

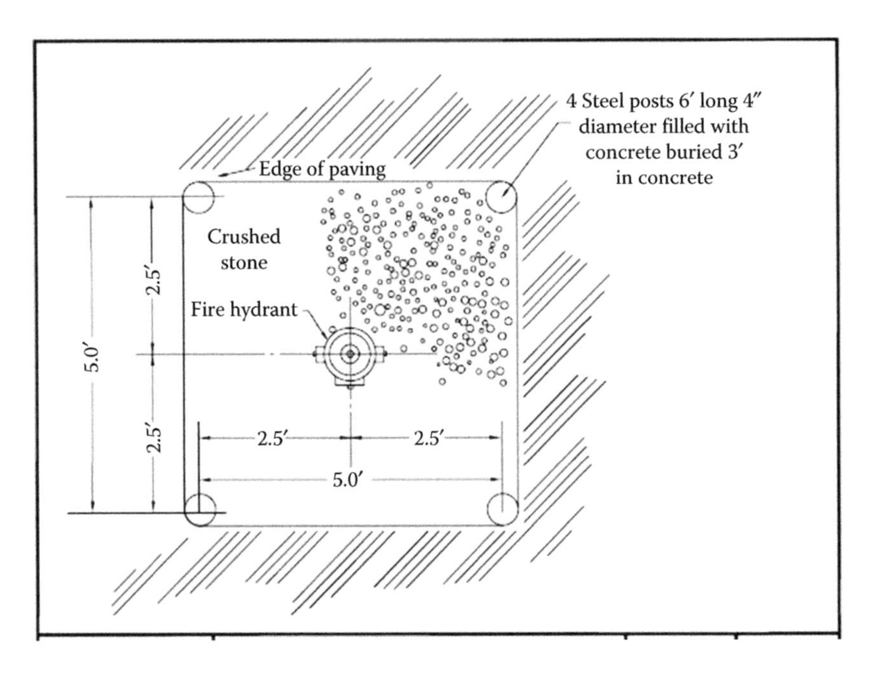

FIGURE 4.12 Fire hydrant protected via bollards. (Courtesy of Fairfax County. With permission.)

FIGURE 4.13 Vehicle blocking access to fire hydrant. (Picture from OSHA: Fire Service Features of Buildings and Fire Protection Systems. www.osha.gov)

locations were random, and no method existed for choosing a location that gave a tactical advantage to responders. Even today, there is no scientific method of how far apart fire hydrants should be spaced or how many are needed. In some jurisdictions, the water authority may be in charge of providing the hydrants.

Three fundamental concepts must be considered when determining fire flow: (1) water needed for the fuel load present, (2) concentration of streams, and (3) accessibility. John R. Freeman, a renowned civil engineer during the mid-1800s, wrote a paper titled "The Arrangement of Hydrants and Water Pipes for the Protection of a City against Fire" in 1892 (NFPA, 2003, pp. 10–67). He conducted many studies covering hose streams that also defined the guidelines for what is used today for the number of fire hydrants based on a building's occupancy. "His work showed how long hose lines reduced the water that can be delivered promptly to the fire" (NFPA, 2003, pp. 10–67). Freeman identified, for fighting fires (to meet the required fire flow), that the focus needs to be on concentration of water rather than distribution. He suggested placing fire hydrants so as to maximize the streams required to supply the site, ranging from 250 feet in between fire hydrants for mercantile/manufacturing districts to 500 feet for residential areas (NFPA, 2003, pp. 10–67).

Similar to this technique is the ISO Grading Schedule method for insurance rate purposes to locate fire hydrants versus maximum flow. "It is a recommended practice that the maximum lineal distance between fire hydrants along streets in congested areas and high fire risk areas with frame buildings and/or high combustible storage (such as lumber), be located 300 feet apart and a maximum of 600 feet, in light residential areas with building separations of over 50 feet" (Harry, 2008, p. 48).

Fire hydrants must be placed at least 40 feet from the building they serve. This is also stated in NFPA 24, and the distance has proven to be adequate for the safety of firefighters who are at the fire hydrants with pumpers, keeping them away from collapsing walls or blown out glass shards/debris from a wall window. However, because of site constraints, some sites may require closer distances. In this case, it is better to locate fire hydrants near the corner of walls. (Figure 4.14).

Wall collapse would occur in the perpendicular direction rather than diagonally. Also, the corners of buildings typically have fewer windows.

So does this mean that every fire hydrant must be at a minimum 40 feet from all buildings on the property? Not quite. The key words mentioned were "40 feet from the building they serve."

In Figure 4.14, the fire hydrant closer to the building is not used to cover that building or that portion of the building because it is less than 40 feet. Instead, the fire hydrants across the compact site can be utilized (Figure 4.15).

When building additions/expansions are proposed, the fire reviewer must ensure that the 40 feet minimum distance to fire hydrants is maintained. If not, the fire hydrant may have to be relocated or additional hydrants must be added to serve the additional portion.

"Other good practices for the installation of fire hydrants calls for at least one fire hydrant at every street intersection, in the middle of long blocks (especially where the needed fire flow exceeds 1,300 gpm), and near the end of long dead-end streams" (Harry, 2008, p. 48). Lastly, fire hydrants that serve FDCs must be placed at a maximum working distance of 100 feet (see Chapter 6).

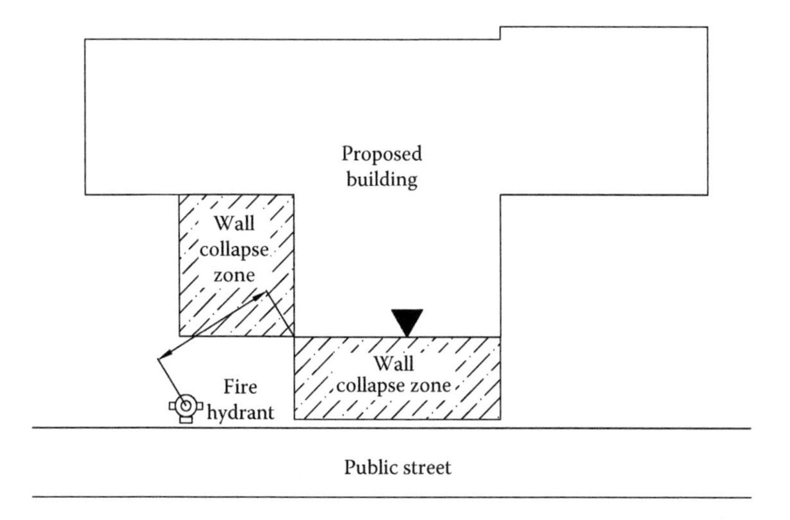

FIGURE 4.14 Fire hydrant placement from building.

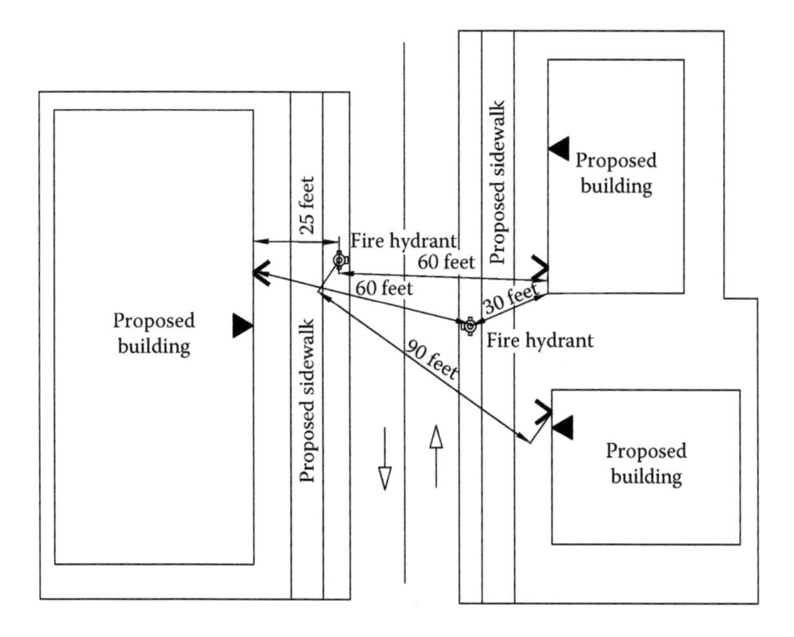

FIGURE 4.15 Criss-cross fire hydrant.

FIRE HYDRANT AVAILABILITY AND ACCESS

On Saturday, November 2, 2014 at 10:53 p.m., more than a dozen firefighters responded to a stable barn fire in Woodstock, Crystal Lake, Illinois (10 minutes after the fire department was called in), and reported heavy smoke and fire coming from a large barn (*Northwest Herald*, 2014). Owing to a lack of fire hydrants in the area, 32 horses were incinerated and the barn was destroyed. The estimated damage was over a million dollars. More than the cost, the horses had been in the family for over 35 years and their loss left an emotional scar.

On March 30, 2015, in Montgomery County, in a non-fire hydrant area, a barn was destroyed by fire with extensive damage to farm structures. Large water tankers had to be brought in to limit the spread of fire (WTOP, 2015).

Freeman's methods have been extracted and adopted by several agencies and AHJ outlining coverage for fire hydrants along a fire department vehicle access way. The IFC has an appendix section dedicated to fire hydrants, where the spacing is based on fire flow. The table below shows the spacing rules used by Fairfax County.

Building Type	**Maximum Distance between Fire Hydrant (feet)**
Industrial buildings and warehouses	250
Schools, day care centers	300
Offices, commercial establishments, church, hospitals, nursing homes	350
Apartments, multifamily dwellings, town houses	350
Single-family dwellings	500

Source: *Fairfax County Public Facilities Manual*, 2011, Section 9.

These measurements indicate what one fire hydrant can cover along the fire access road, but not over buildings, or fences or curbs. Furthermore, it can be seen from the table that, depending on the building type, the spacing changes. This is because of the changes in fire fuel load present depending upon the occupancy, which thereby increases the number of streams required to counter the generated heat. Many people get the idea that fire hydrants have to cover all around the building perimeter. This is not exactly true. Remember that fire trucks connect supply hoses to fire hydrants; so the distance is based on how far the truck can travel from a fire hydrant. Furthermore, there may not be adequate pressure at the fire hydrant to achieve proper hose streams for handheld hose lines when connected directly without a pumper. As we will see in Chapter 7, it is imperative to have fire truck access to the buildings so that the trucks can pull up close to the buildings and achieve suppression along the perimeter.

Figure 4.16 shows coverage of fire hydrants along the path of fire truck travel with hose lay. Fire truck access is required to be within 150 feet of all portions of the exterior wall for the story at grade.

Many engineers make an easy observation and make a circle around the fire hydrant to show coverage. This is wrong. The hose is not laid as the crow flies, there is no truck access over uneven terrain, and there are also many obstructions, such as trees, ponds, etc. (Figure 4.17).

At first sight, looking at the radius, it appears that coverage is met for this single-family dwelling unit. However, if we measure this distance along the fire department vehicle access way (as hose is laid), we see that coverage is not met.

Though both frontlay and backlay hose methods are used, fire hydrants should be located along the path of the fire truck travel. This means they should be anywhere from the fire station to the building on fire (within the fire hydrant coverage

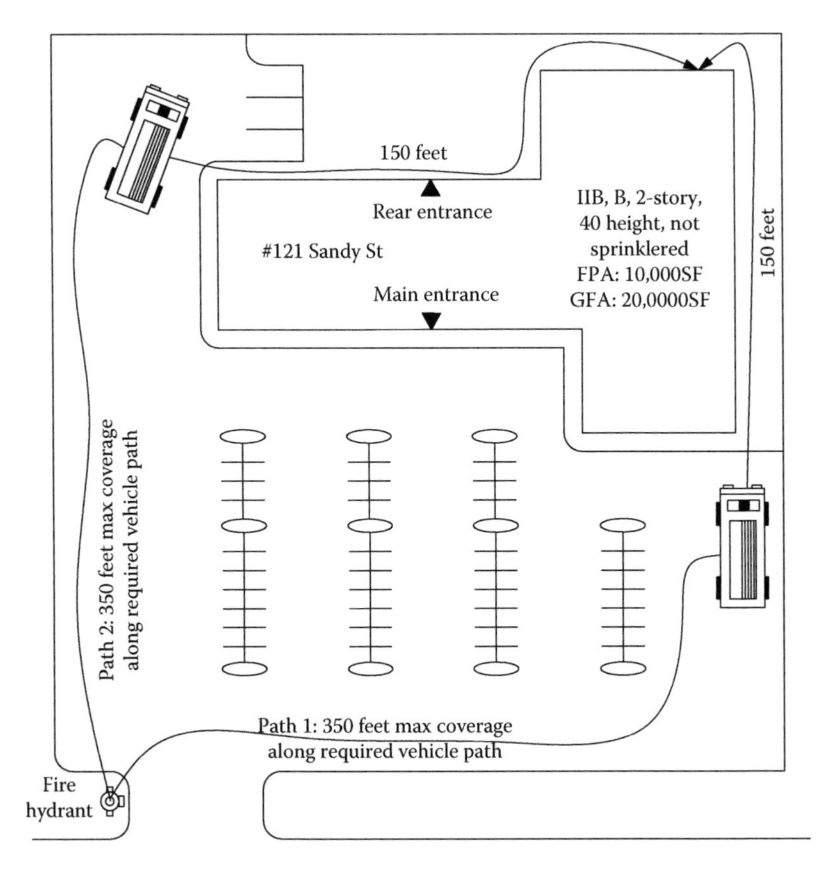

FIGURE 4.16 Fire hydrant coverage.

distance). Trucks should not pass the building on fire. For instance, hydrants located on dead ends should be used to serve the lots at the dead end only. This is because the dead end may have insufficient turnaround space and it will increase response time to go down to the end and return. Remember, the purpose of a fire site review is to make it so that the response can be executed efficiently. The IFC defines "fire apparatus access road" as a "road that provides fire apparatus access from a fire station to a facility, building or portion thereof" (ICC, 2009). The key terms to note are "from a fire station to a facility, building," *not* past it. Existing fire hydrants should be utilized wherever possible. If there are multiple entry points to the site from an accessible road, any hydrant along that path should be considered acceptable to provide the fire flow (Figure 4.18).

Along major intersections and roads, fire hydrants should be provided on both sides of the street. This will help prevent a major road shutdown as the hose will not have to be laid across the street causing traffic congestion, which would prevent other response units from accessing the site quickly. Therefore it is best to have fire hydrants preferably at the site entrance and not along public/main roads whenever possible. When placing hydrants, consider how the hose will be laid along the street.

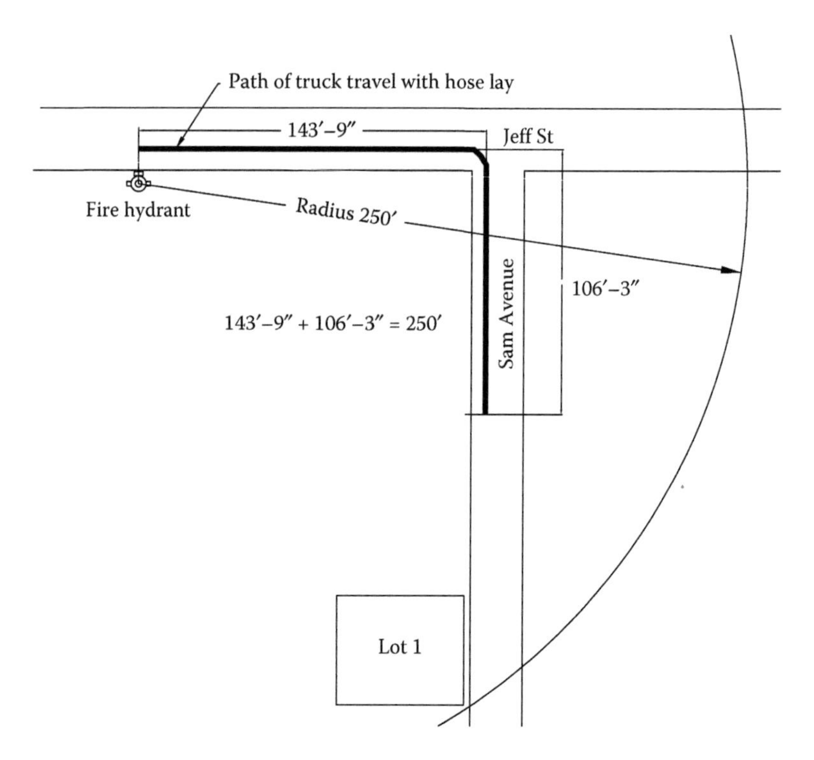

FIGURE 4.17 Fire hydrant coverage.

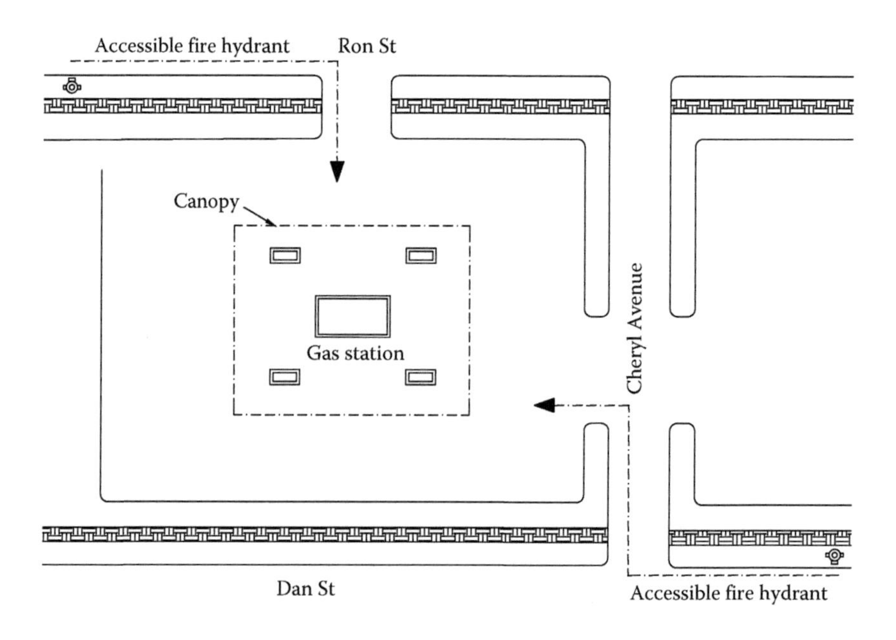

FIGURE 4.18 Fire hydrant coverage.

Similar to FDCs, the threads on a fire hydrant must be compatible with the fire truck hose. National Standard threads (NST) are widely used. Check NFPA 1996. Adapters can be used where the hoses do not match. Private fire hydrants should be maintained and tested in line with NFPA 25. NFPA 24 and 291 are two useful standards with respect to fire hydrants.

REFERENCES

Harry, H. 2008. *Water Supply Systems and Evaluation Methods—Volume I: Water Supply System*. U.S. Fire Administration FEMA. http://www.usfa.fema.gov/downloads/pdf/publications/Water_Supply_Systems_Volume_I.pdf

International Code Council (ICC). 2009. *International Fire Code 2009*. Country Club Hills, IL: International Code Council, Inc.

NFPA. 1996. *NFPA 1963: Standard for Fire Hose Connections*.

NFPA. 2003. *Fire Protection Handbook*. 19th edn. Quincy, MA: National Fire Protection Association, Inc.

NFPA. 2006a. *NFPA 1142 Standard on Water Supplies for Suburban and Rural Fire Fighting*. Quincy, MA: National Fire Protection Association, Inc.

Northwest Herald. 2014. Firefighters believe 32 horses died in stable fire between Woodstock, Crystal Lake. Retrieved on November 24, 2014 from http://www.nwherald.com/2014/11/23/firefighters-believe-32-horses-died-in-stable-fire-between-woodstock-crystal-lake/au7xyp/

Sturzenbecker, M. J., Adams, B., and Burnside, E. (Eds.). 2012. *Fire Detection and Suppression Systems*. 4th ed. Stillwater, OK: Fire Protection Publications.

WTOP. 2015. Nearly 2 dozen pigs killed in Montgomery County barn fire. Retrieved on March 30, 2015 from http://wtop.com/montgomery-county/2015/03/nearly-2-dozen-pigs-killed-in-montgomery-county-barn-fire/

5 Underground Fire Lines

A fire plan reviewer examining site plans should be aware of most, if not all, of the general fire safety features of the building. These are the fire suppression sprinklers systems and their standards (NFPA 13, NFPA 13R, NFPA 13D), the fire alarms, and even life safety features, such as fire exits. After it is determined through the building code that a building is required to have an automatic fire sprinkler system or standpipe, certain key elements of this system must be shown on the site plan. These include Siamese connections, fire hydrants serving the Siamese connections, exterior location of alarm bell, and underground fire lines. Fire lines are underground water-carrying pipes that provide water from the source to the supply inside the building for the built-in fire suppression system (i.e., the sprinkler system) (Figure 5.1).

For sites where large water mains are not present to provide adequate water, other sources such as tanks, ponds, and retention can be used. The utility plan shows the locations of all utilities on and around the site, including fire lines, water lines, underground duct banks for electrical services, gas lines, storm water drains, sanitation, overhead power, etc. In cases where large public water mains are available to the site, an underground fire line can be tapped into the public water distribution (Figure 5.2).

Typically, after the water easement, a fire line is privately owned and maintained. This private fire line must adhere to standards such as NFPA 24 Standard for the Installation of Private Fire Service Mains and Their Appurtenances. In fact, NFPA 24 can be used for all private fire services that include fire hydrants. For a site plan review, factors to consider for fire lines are their depth, their size, whether it is a combination of both fire and domestic supply, valves on the line, and clearances to other utilities. A listing of pipes and materials, bends, and thrust blocks are beyond the scope of site plans, so formal shop drawings should be submitted separately after site plan approval.

The size of a fire line should not be less than 6 inches, unless determined by calculations based on NFPA 13 requirements. The NFPA 13D and 13R sprinkler systems for single-family dwelling units, townhouses, and small residential buildings will have a smaller system demand, so the fire lines will be sized smaller. The site plan reviewer must be aware of the danger of undersized fire lines. If the reviewer feels that the size does not appear to be sufficient, he/she must ask the engineer to show calculations to justify the size used.

Although it can be cost effective to use one line to supply both, normally the fire line and domestic lines are branched off separately from the street main. Combination lines help provide automatic monitoring, as turning off the water supply for the fire line will turn off the building's potable water supply letting the owner know fire sprinkler system is shut off. The combination line must not only meet all fire safety requirements (inspection and testing), but also maintain potable water until the lines separate. It is good practice to split the fire and domestic lines outside the building, with a valve only on the domestic line. There should be no valve on the

FIGURE 5.1 A long run of underground pipe during construction.

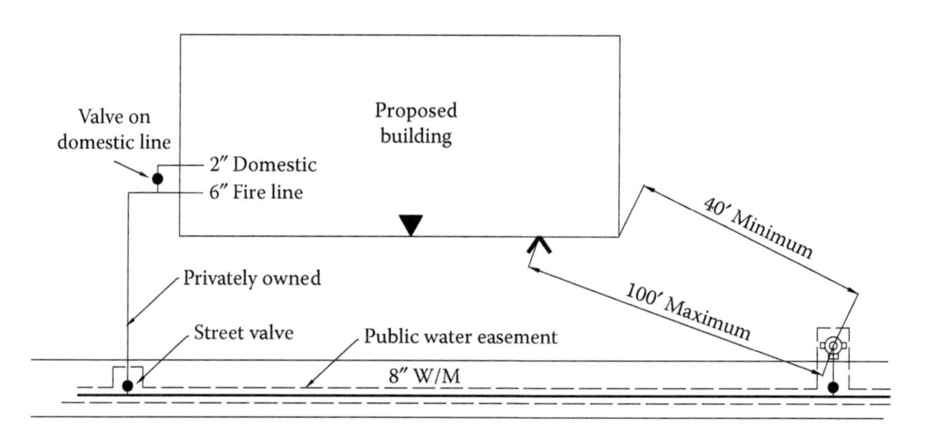

FIGURE 5.2 Separate domestic and fire line.

fire line except for the street valve and the control valve inside the building, as shown in Figure 5.2. This will eliminate any accidental shutoff of the fire line that might occur after any repair work.

TALL BUILDINGS

There are some building codes that require a secondary water supply for localities in seismic areas. Furthermore, when the fire department pumper is unable to meet the standpipe system demand, a secondary means of water supply is required. This is mentioned in *NFPA 14: Standard for Installation of Standpipes and Hose Systems* when more than one standpipe zone is established due to pressure restriction requirements. The secondary water supply can be in the form of an elevated water storage tank on the roof of the building, or a fire pump, or another reliable water source supplying both pressure and flow acceptable to the AHJ. The former (the roof tank) will require significant design changes to the building structure and increase the cost. Buildings over 300 feet must be evaluated for this, however the reviewer must be aware of the local pumper discharge pressure and city street pressure, both at the required flows, in order to determine whether or not the system demand will be met (Figure 5.3).

City pressure + fire pumper pressure must be greater than or equal to the system pressure at the top outlet (100 psi) + elevation loss + friction loss in hose from fire hydrant to pumper + friction loss of hose from pumper to FDC + friction loss in pipe and fittings from FDC to outlet at the required flow.

Jurisdictions that have adopted the ICC codes need to be aware of duel fire line requirements. A section was introduced in 2006 that requires that every high-rise building requiring a pump needs to have two fire lines. Almost always, a high-rise building will require a pump due to high elevation, unless the street pressure is extremely good. High-rise buildings present a significant risk. On the basis of ICC commentary, an unnoticed broken underground fire line can leave the entire system inoperable. For this reason, two fire lines are required from different water supply mains. If they cannot be from different systems, they must be valved in such a way that a disruption in the water supply from one side can be isolated and still supply the system demand from the other fire line (ICC, 2011).

Figure 5.4 shows two fire lines from different water mains. Figure 5.5 shows two fire lines from the same water main. The street valve between the two fire lines

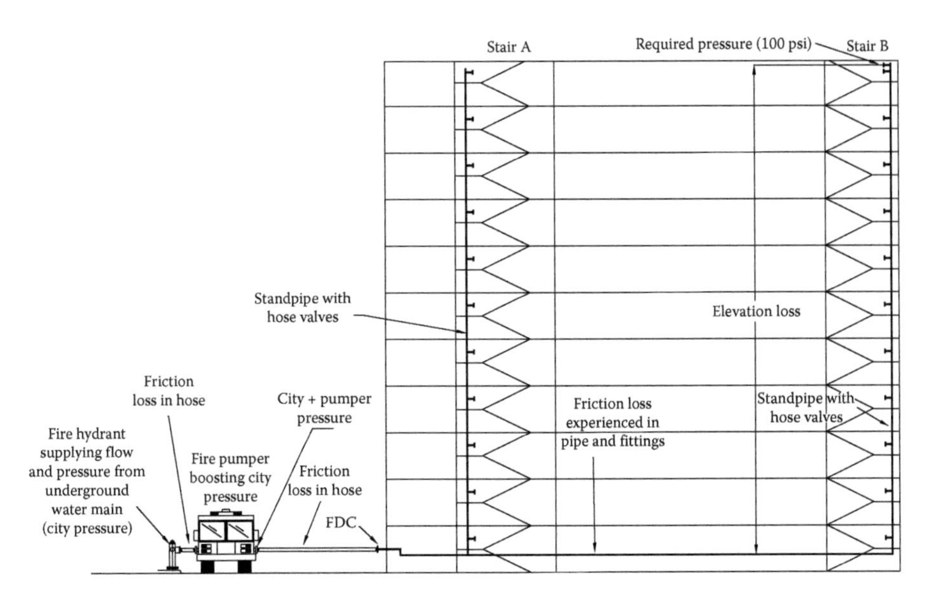

FIGURE 5.3 Fire pumper pumping from fire hydrant to standpipe.

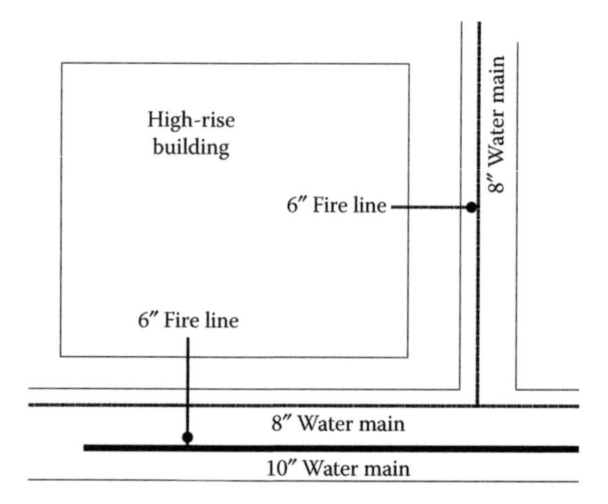

FIGURE 5.4 Fire lines from separate water mains.

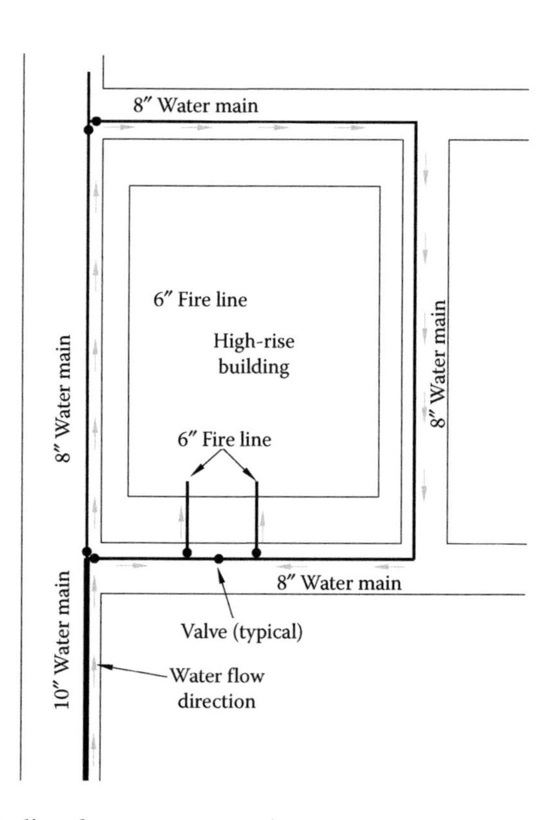

FIGURE 5.5 Fire lines from same water main.

can be shut off and isolate the portion on the 8-inch loop if a break occurs, while still having water flow from the other side. Note that a possibility does exist for the 10-inch main also to break and render the loop ineffective. However, since the 10-inch main is a major supply line, it will likely be repaired faster, during which time the building should be on fire watch.

CASE STUDY: NFPA 13R SPRINKLER SYSTEM

A site plan shows a group of town houses four-stories, 49 feet in height, of type VB (wood) construction. In line with the IBC, an NFPA 13R system is required, and is indicated on the site plan to be installed. The plan shows a 1-inch fire and domestic combo line with a 3/4-inch meter (Figure 5.6). Does this seem acceptable?

Analysis: The first thing the fire reviewer must do is to verify the height and area of the building in order to make sure that an NFPA 13R sprinkler system is the correct system for this building. In line with the IBC, a 13D system does not allow for an increase in height or the number of stories, so either an NFPA 13 or 13R sprinkler system must be used. This building also does not fall under the International Residential Code (IRC) as it is more than three stories. A 13R system can require up to a four sprinkler head hydraulic calculation. At a minimum operating pressure of 7 psi, using a 5.6 K head, this will result in 14.8 gpm ($Q = k\sqrt{P}$, where Q is flow, k is the k-factor, and P is pressure). This means that the total sprinkler demand is 59.2 gpm (4 × 14.8 gpm). Often, listed residential heads are used, which allow for an increase in coverage area. This results in the installation of fewer heads, but then usually both flow and pressure exceed 14.8 gpm at 7 psi; see the sprinkler manufacturer cut sheet to attain the correct characteristics. NFPA 13R requires that the domestic demand should also be added to the sprinkler system demand. This is because, during the sprinkler head operation, other areas in the house might be utilizing water, that is, someone taking a shower and unaware of the fire developing

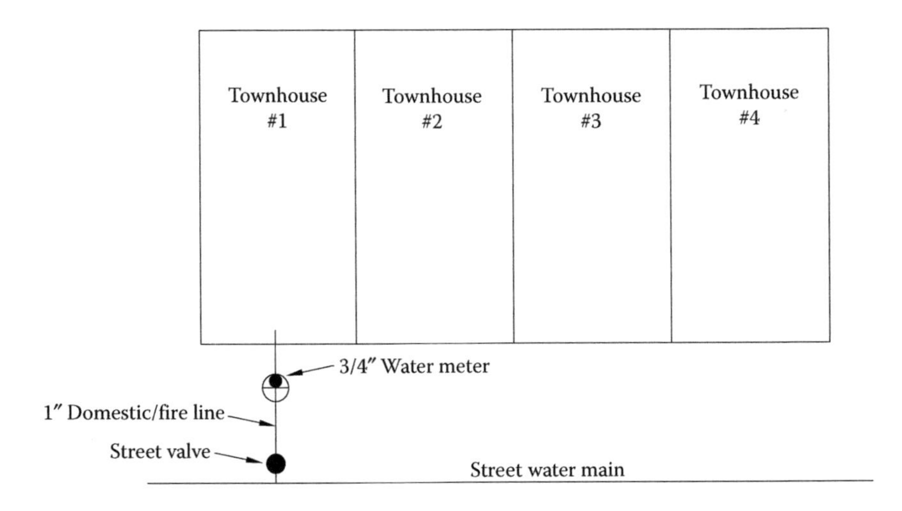

FIGURE 5.6 Example of townhouse with common fire and domestic water line.

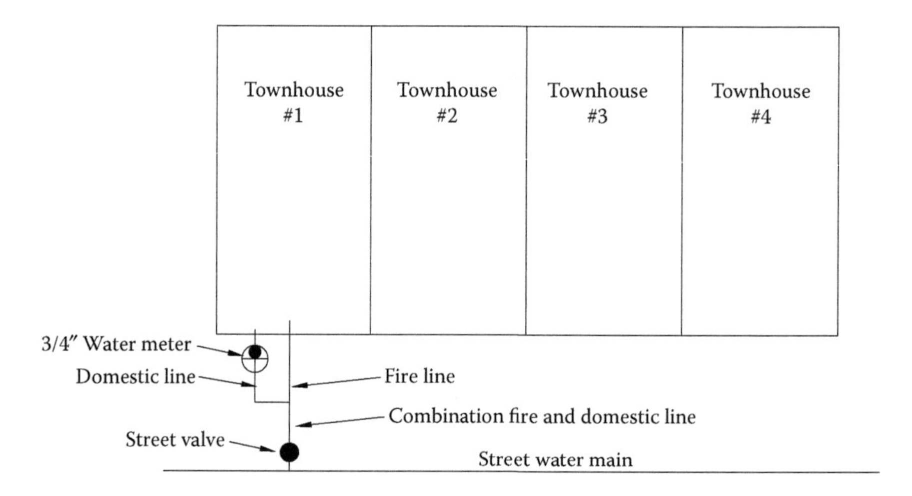

FIGURE 5.7 Townhouse with common fire and domestic water line where split occurs outside.

in the basement. According to NFPA 13R, this can range from approximately 26 to 45 gpm. Therefore, the final sprinkler system demand can reach somewhere between 85.2 and 105.2 gpm.

Problem: On the basis of this flow, there will be significant friction loss not only through the 1-inch combination fire and domestic line, but also through the 3/4-inch meter. This may require a pump to be added to boost the pressure. However, both NFPA 13R and NFPA 13 sprinkler systems require a listed NFPA 20 fire pump, which will not be ideal in terms of either cost or maintenance. Furthermore, upsizing the combo line and the meter can also increase costs significantly.

Solution: The best approach is to have a separate fire line and a separate domestic water main, or have a split occur just outside the building with no meter on the fire line. Also, remember that a fire line cannot run under the building unless special considerations are taken into account, according to NFPA 24/NFPA 13. It should come up immediately (typically within 5 feet), as soon as it enters the building (Figure 5.7).

With this layout, the flow has to be calculated only in the fire line plus the domestic line at the split. The fire line and the combo line can be increased in size to reduce friction loss.

Commercial buildings with an NFPA 13 sprinkler system follow the same design. Typically, no meter is present directly on the fire line, although the water authority may require one to be installed for monitoring. Instead of having a meter right on the fire line (due to cost and friction loss), a fire backflow preventer (BFP) assembly with a factory-fitted meter can be utilized.

PROFILES

We touched a little bit on profiles in Chapter 1. The profile of a fire line shows the run of pipe throughout the site, along specific station markers that help identify its location. Profiles for fire lines are generally found on the water line profile sheet (Figure 5.8).

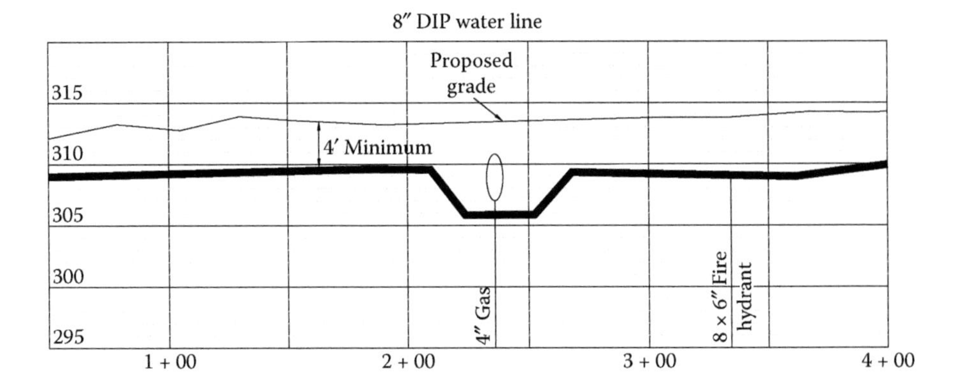

FIGURE 5.8 Example of water/fire line.

Typically, the scale on these profiles is 1:5. The fire site plan reviewer should examine the following items:

1. The size of the pipe and verify that it matches that on the plan view.
2. The material it is made of, for example, ductile iron (DI).
3. The depth of the pipe: In areas where freezing is of concern, the top of the pipe should be buried 1 foot below the frost line. NFPA 24 and NFPA 13 both provide a map of North America indicating the minimum burial depth of pipes for fire protection systems. Consideration should also be given to protecting the pipe from mechanical damage. The maximum cover of the pipe should take into account the ease of digging to make repairs. It is good practice to bury the pipe not more than 7.5 feet. One thing to note is that the water in the fire line is stagnant (until operation or testing), and if the fire line is in combination with a domestic line, then after the takeoff (at the split), the water in the fire line will still be stagnant. For that reason, it is important to make sure that the fire line portion (beyond the domestic) is below the frost line. Figure 5.9 illustrates this.
4. Clearance between other utilities: A physical clearance should be maintained between the fire line and other utilities. Where the fire line is taken off a public water distribution system, similar clearance should be followed. The Fairfax County Fire Marshal's Office—Engineering Plan Review has the following guidelines:

Fire line–storm sewer crossings (at 90° more or less): If the fire line is above storm, a 6-inch clearance is required. If the fire line is below storm, a 12-inch clearance is required.

Fire line–sanitary sewer crossing (90° more or less): If the fire line is above a sewer, an 18-inch clearance is required.

Fire line–gas main crossings (90° more or less): 12-inch clearance above or below.

Fire line–electrical service entrance conductor: 12 inches. The fire line is below electrical service, and the maximum depth of electrical service is no greater than 36 inches.

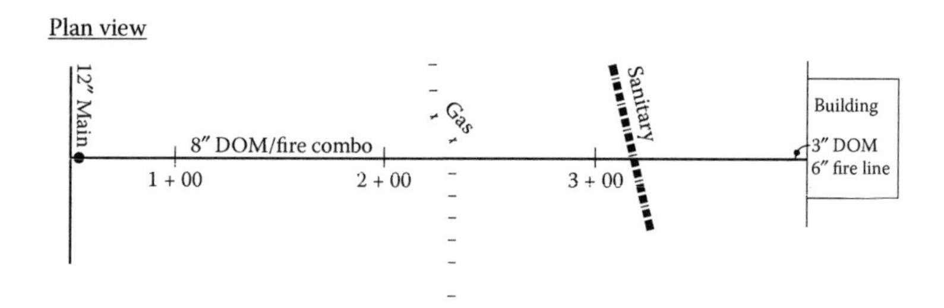

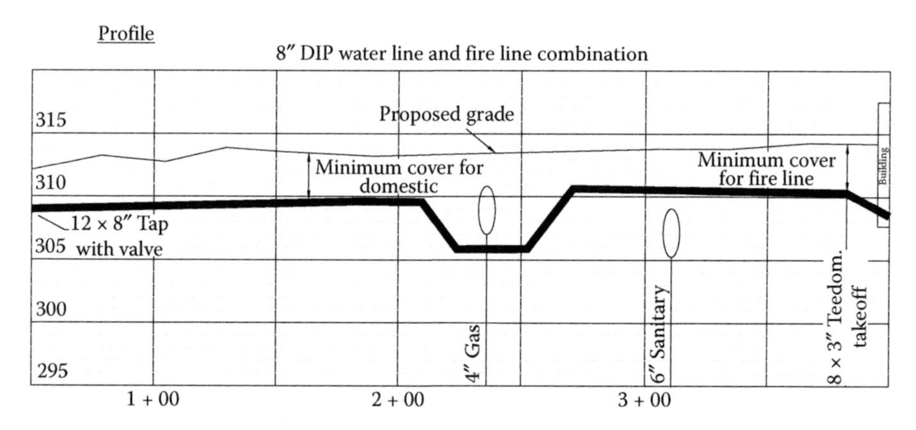

FIGURE 5.9 Profile of water/fire line showing crossing of other utilities that is, gas and sanitary.

The minimum depth of electrical service is 24 inches. All other electrical underground wiring (e.g., cable TV, fiber, etc.) is to be treated as same as electrical service.

1. Other utilities are not to run in the fire line trench.
2. At crossings, where the other utility is above, the intervening fill should be compacted granular material, 90% Standard Proctor, AASHTO T-99) (per DIPRA, installation guide for DIP).

Summary Table of Crossing Clearances

Fire line to storm	6 inches if fire line is above, 12 inches if fire line is below storm
Fire line to sanitary sewer	18 inches, sanitary must be below fire line unless met with state health requirement
Fire line to gas main	12 inches, above or below
Fire line to electrical service	12 inches, electrical above, fire line below

These were developed by David Thomas in conjunction with the Fairfax Water Authority. Check with the water service provider in your area and also State requirements. Contact other utility agencies to determine their clearance requirements. Remember that any problem that can be corrected on a site plan will avoid difficulties in the field.

In warm climates, the BFP can be located outside the building. The reviewer must ensure that the location is appropriate and protected from vehicle impact and damage.

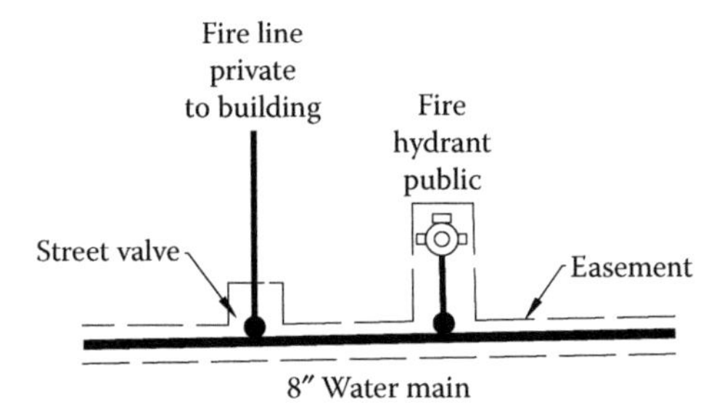

FIGURE 5.10 Water main showing easement.

EASEMENT

On a site plan, an easement is a dedicated pathway for a utility provider to use on an owned property; in other words, it is a right-of-way. Water lines, power cables, fiber optics, and gas lines all have easements. You can typically see the easement around utilities on a site plan or a utility plan. They are represented by dashed lines around the utility (Figure 5.10).

With an established easement, no other utility has the right to run inside another easement, although crossing is allowed. Private fire lines do not have an easement, nor are there any restrictions as to how close other utilities can run to them (except that they have to meet their own clearance requirements). It is important to determine who has the authority over the fire line. In some jurisdictions, the water authority may be responsible. In a case where it is privately maintained, no other pipe should run in the fire line trench. Lastly, in line with NFPA 24, when post indicator valves are used to control water supplies, they must be located not less than 40 feet from the building. The reviewer must confirm that the valves are located in an accessible location, away from parking lots and curbs to avoid damages. Vehicle impact protection (bollards) should be shown on plans where needed.

SITE RENOVATIONS

When a site plan is altered, it is necessary to look at the changes to make sure that existing fire line(s) meet the code. An example of such an instance is when a utility is added, such as an underground storm water line. Clearances must be maintained between the fire lines and other utilities, as mentioned above. Even small additions to a building must be reviewed to ensure that a violation has not occurred.

CASE STUDY

An existing office building owner wants a small new addition of 200 feet^2 to his building for office space. The building is fully sprinklered, and the new addition will also be sprinklered. (Figure 5.11).

At first sight, everything appears to be code compliant. However, after looking at the existing plan, we can see that the sketch does not show the location of the fire

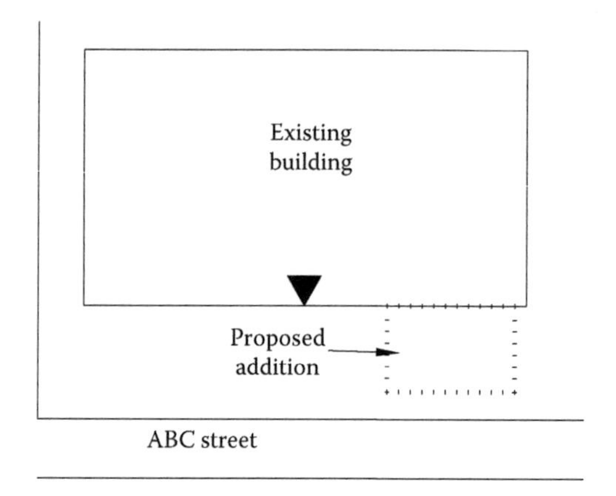

FIGURE 5.11 Existing site showing no utilities.

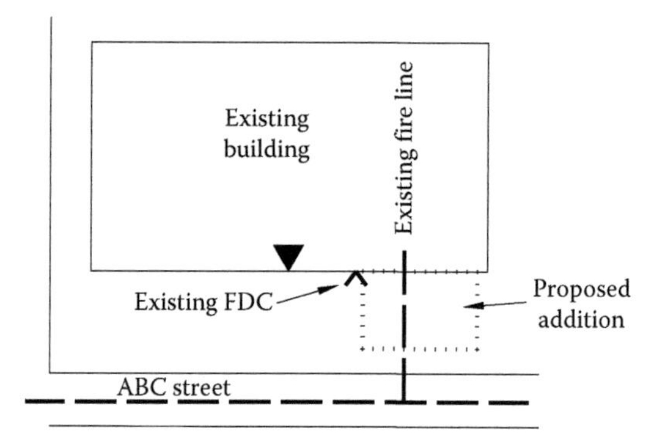

FIGURE 5.12 Existing site showing utilities (fire line and FDC).

line or FDC. The reviewer must request a plan showing the fire line, fire lanes, and FDC to ensure that they have not been compromised.

In Figure 5.12, the existing fire line is running under the building and the existing FDC location is in conflict with the addition. Fire lines are not allowed to run under a building according to NFPA 24 (unless special conditions are met). If the fire line requires repair, then the new addition would have to be excavated, resulting in very high cost. Weight of foundation and slab can cause stress on fire line resulting in damage or dislodging. The owner can choose to relocate the fire line and FDC, which would then require retesting.

REFERENCE

International Code Council (ICC). 2011. *2012 International Building Code and Commentary.* Country Club Hills, IL: International Code Council, Inc.

6 Fire Department Connections

The terms *Siamese connection* and *fire department connection* mean the same thing and are used interchangeably throughout this text. FDCs tie directly to the building fire suppression system or standpipe. For automatic suppression systems (fire sprinklers, water mist, foam system), the purpose of an FDC is supplementary, meaning that it serves as an additional supply. For other suppression systems, such as a foam chamber serving large flammable tanks, the FDC can be the primary means to provide water and pressure. We will concentrate on automatic sprinkler systems. A fire pumper can use the FDC to boost the pressure of an operating sprinkler system and standpipes (Figure 6.1).

Note that the designed automatic sprinkler system according to NFPA standards must be able to control the fire without intervention by fire personnel and a fire pumper. In simple terms, the FDC is used by the fire department when the incident commander feels that the fire is not contained. Typically, when a fire call is made to the building, the responding units will hook up the fire pumper to the FDC so that it is ready to be used if needed. The pressure provided by the pumper does not add to the pressure supplied to the sprinkler system via the underground fire line. Pressure balances to the higher pressure. This is similar in concept to two common points on a branch line of a sprinkler hydraulic calculation. For cases where the building is required to have a standpipe (governed by building codes), the FDC allows firefighters to pressurize the standpipe, thereby also increasing flow and use the hose valves at a desired pressure on a particular floor level for interior or roof manual firefighting using hand-held hose lines. Where the buildings are taller, beyond the fire department pumping capacity, an on-site fire pump must be provided to meet the demand of the standpipe pressure. Most likely there will already be a fire pump for the sprinkler system, so it can be sized larger to meet the standpipe demand as well.

In *Fire Service Features of Buildings and Fire Protection Systems*, the Occupational Safety and Health Administration (OSHA) states, "any deficiency related to the FDC can cause delays in fire suppression, and therefore a decrease in the safety of both firefighters and building occupants" (OSHA, 2006). So what is the function of an FDC? A fire pumper truck makes a suction connection to a fire hydrant that is serving the FDC. The discharge connection from the pumper is made to the FDC. The fire truck takes the water from the hydrant or the source supplying water, boosts the pressure via the pumper, and supplies the flow through the FDC that is fastened directly to the standpipes and the suppression system. NFPA 1901 states the pressure requirements that must be supplied by the fire pumper. Pumpers rated for 3,000 gpm or less must deliver the following net pressures: 100% of rated capacity at 150 psi, 70% at 200 psi, and 50% at 250 psi.

FIGURE 6.1 Wall-mounted Y-type FDC. (Courtesy of Fairfax County. With permission.)

Depending on the pressure available from the fire hydrant, a flow of 300 psi can be achieved at the FDC inlet. A key item to note is that the fire pumper does not create water. It has the ability to increase flow, which basically is the principle of increase velocity (speed) of water through the outlet/nozzle by means of increasing pressure via the fire pump on pumper. There must be water present at the source otherwise the fire pumper will create a negative pressure at the hydrant (sucking water) which could collapse the water distribution system and contaminate potable water system. In the event of the primary water supply (via the fire line) being inoperational, or the system being unable to control the fire—due to impairment in the fire line or too many heads overflowing—the FDC can function as an alternative water supply.

The most common FDCs comprise two 2½-inch inlet connections. One 2½-inch connection can deliver 250 gpm without experiencing significant vibration or pressure loss in the pipe and hose. This is also stated in NFPA 14. Depending on the system demand, there may be a number of inlets for the FDC.

To meet the standpipe demand of 1,000 gpm for a fully sprinklered building, at least four 2½-inch inlets are required. For 1,250 gpm, five inlets are required. For buildings that only have sprinkler systems, the number of inlets depends on the flow required by the most demanding system.

Both the IBC and NFPA 13 provide guidance on the location of a Siamese connection. FDCs on the site plan must be clearly marked. Although not esthetically preferred, the ideal location is the street front, the address side of the building, preferably near the main entrance, so that it is visible to the oncoming fire truck dispatched to the site. This is because the first-arriving emergency vehicle tends to arrive at the main entrance, which allows personnel to carry stretchers, communicate with the people exiting, get to the annunciator panel (usually located in the main vestibule if the building has a fire alarm system), etc. Sufficient clearance should be made around the FDC so that it is visible. It should not be located near nooks. The firefighters must have adequate working space and must be able to make a hose connection from the pumper easily (Figure 6.2).

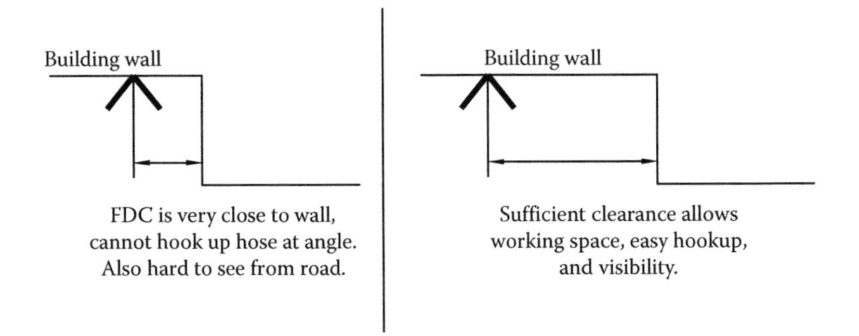

FIGURE 6.2 FDC location near wall nook.

PLAN VIEW

Both IBC and the IFC indicate a clearance requirement of 3 feet to be maintained front and sides and 6.5 feet above (Figure 6.3). Examine areas where you have bike racks, shopping carts, gas cylinders, gas meters, loading docks, outdoor storage display, and car parking lots with reference to FDC locations. Keep FDCs clear of fences, bushes, trees, and other obstructions.

Figure 6.4 shows a wall-mounted FDC. A few years from now, the plants and shrubs around will grow to obstruct the location and visibility. It is best to err on the side of caution and relocate the FDC at design stage or clear all planting. Confirm this by looking at the landscape plan (Figure 6.5).

A flush-mounted FDC can also be used, but clearance should still be maintained for visibility (Figure 6.6).

FIGURE 6.3 The hose outlet on the FDC is too close to the wall. A hose connection cannot be made to charge the standpipe/fire suppression system. (Courtesy of OSHA.)

FIGURE 6.4 FDC location close to planting. Over time bushes may obstruct FDC. (Courtesy of Kris Lacy, Fairfax County. With permission.)

The reviewer should also note any retaining walls near the FDC, as they can act as a barrier. If for any reason the FDC is not visible, signs can be made to indicate its location. The IFC and the NFPA outline some methods for appropriate signs. When there are multiple FDCs serving different buildings or systems in a building, signs/labels should be provided at the FDC to indicate which area it serves. It is good practice to keep the FDC away from doors, windows, and vents, which in case of fire can project flames, heat, and smoke.

FIGURE 6.5 Picture showing both the FDC and the pump test header. The truck restricts the fire department from utilizing the FDC in the event of a fire in the building. (Courtesy of Fairfax County. With permission.)

FIGURE 6.6 Flush-to-wall FDC.

Figure 6.7 indicates how charged (pressured) lines to the FDC from a pumper can obstruct an exit. The reviewer should evaluate the FDC location with respect to exits. A good practice is to place FDCs 10 feet away from openings.

There may be times when multiple FDCs are needed for a building. An example of this is an open parking garage with manual dry standpipes for five stories above grade, with a dry pipe sprinkler system two stories below grade. For this building, at least two separate FDCs are required, one serving the manual dry standpipes and the other serving the dry sprinkler system. Both FDCs should be located at street front, address side of the building, and have a sign indicating the area of the building they serve. As mentioned in Chapter 1, a garage may not have an address. It is good design to provide an additional FDC (interconnected with the primary) at a remote end, usually on the second entry side to the garage, for better accessibility, especially for

FIGURE 6.7 Pressurized FDC hose blocking exit. (Courtesy of OSHA: Fire Service Features of Buildings and Fire Protection Systems. www.osha.gov)

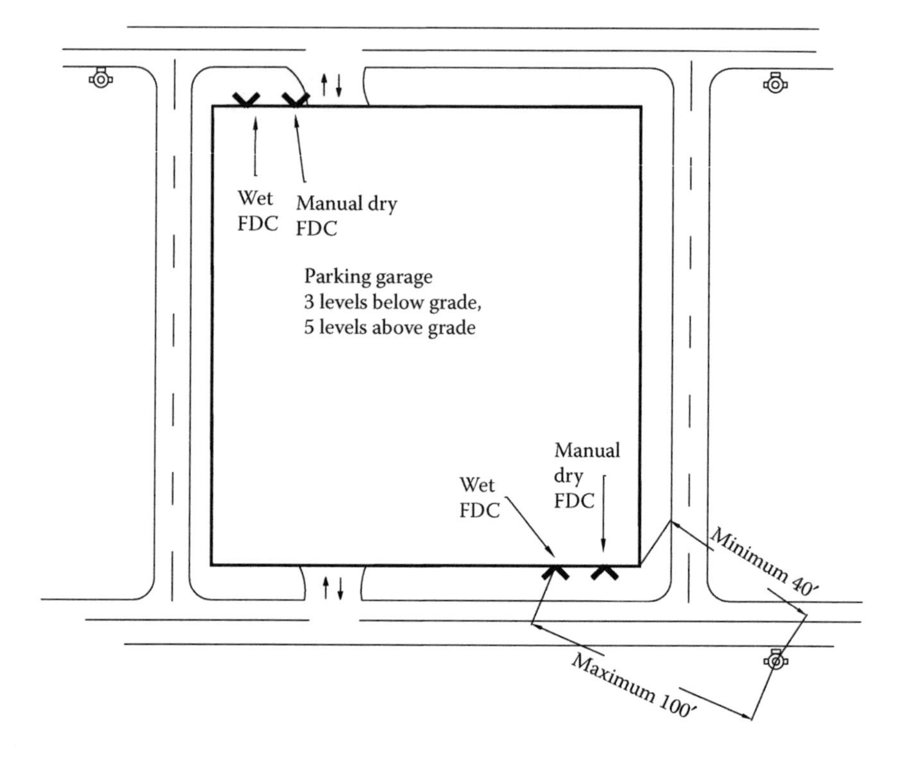

FIGURE 6.8 Garage FDCs on two sides.

big garages. In such a case, it is best to locate it near corners, away from the garage entry, as cars may be trying to get out (Figure 6.8).

Where a structure is of pedestal design with multiple buildings on top of a podium with different addresses, having the FDC at the right location is of crucial importance (Figure 6.9).

FDCs for each building are located on the building that they serve. Having the FDCs located at one location for complex structures can make it difficult for responders to figure out which one to pressurize. Finally, NFPA 13 gives mounting heights for FDCs from 18 inches to a maximum of 4 feet above finished grade.

FIRE HYDRANT SERVING A SIAMESE CONNECTION

At least one fire hydrant must be made available to serve the FDC. The distance to this serving fire hydrant is based on the time it will take for one operator to connect the hose from the pumper to the fire hydrant and another hose from the discharge side of the pumper to the FDC. Furthermore, the hose lay should limit obstruction on the road for other vehicles and personnel entering and leaving the site. Recall from the Analysis section, at the beginning of this book, the Rockville apartment fire, where people occupying the neighborhood adjacent to the building on fire were trapped and unable to get out due to hose lay blocking the streets. While 40 feet is

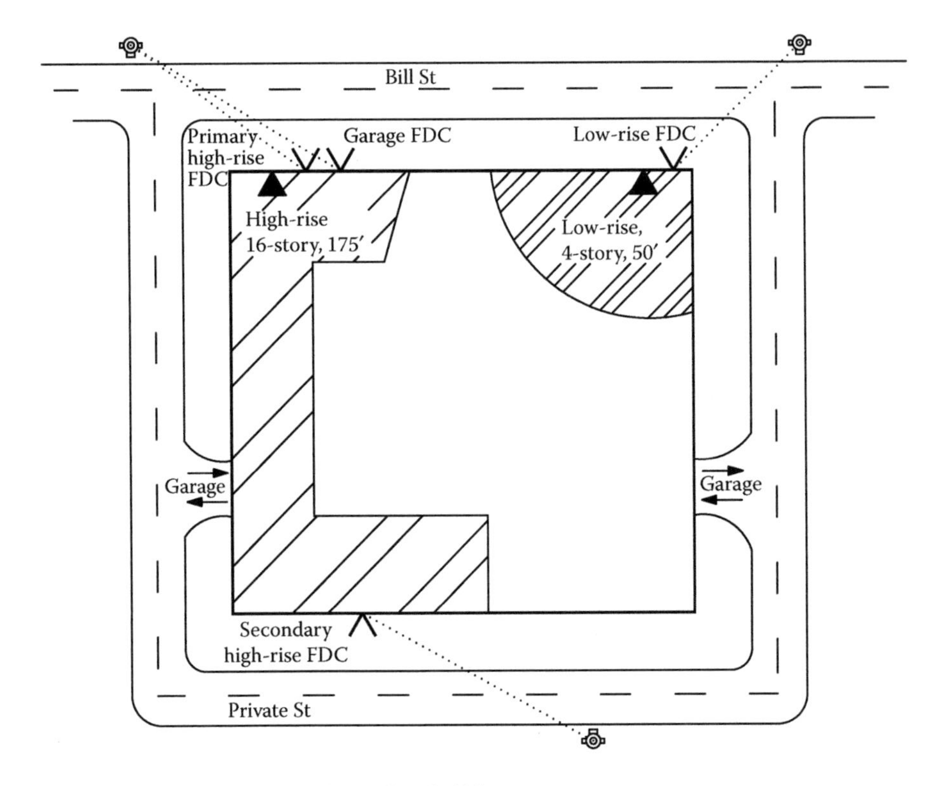

FIGURE 6.9 FDC location for podium buildings.

the minimum distance a fire hydrant must be away from the building, a maximum of 100 feet is considered the furthest acceptable distance a fire hydrant should be from the FDC it serves. Basically, this means the fire hydrant serving an FDC should be located anywhere from 40 to 100 feet away from it. This allows for a fast connection from the pumper to the fire hydrant (Figure 6.10).

The distance can be measured over parking and other obstructions. This is because the hose is not connected from the fire hydrant to the FDC, but rather from the fire hydrant to the pumper, and from the pumper to the FDC. Another consideration that should be taken into account, as mentioned in the IFC, is how the hose is laid from the FDC to the hydrant via the pumper when a single access point is provided to the site. The hose lay should not prevent other emergency vehicles from passing when it is charged. For this reason, where new hydrants are being proposed to serve FDCs, consider how the hose would be laid. If possible, use a free-standing FDC. Figure 6.11 shows how the pumper can connect to the FDC and the fire hydrant all on the same side of the street, thus allowing other vehicles to pass by.

FDCs FOR HIGH-RISE BUILDINGS

According to NFPA 5000, a high-rise "is a building where the floor of an occupiable story is greater than 75 feet above the lowest level of fire department vehicle access"

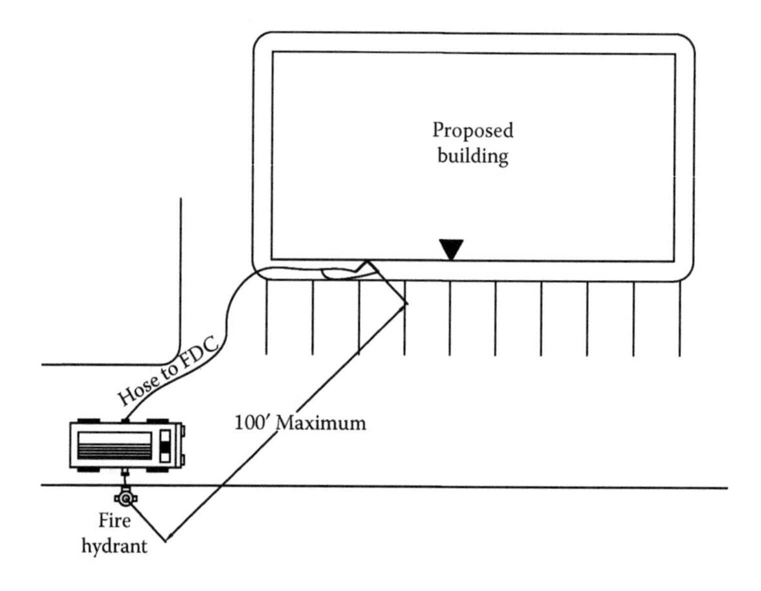

FIGURE 6.10 FDC distance to fire hydrant.

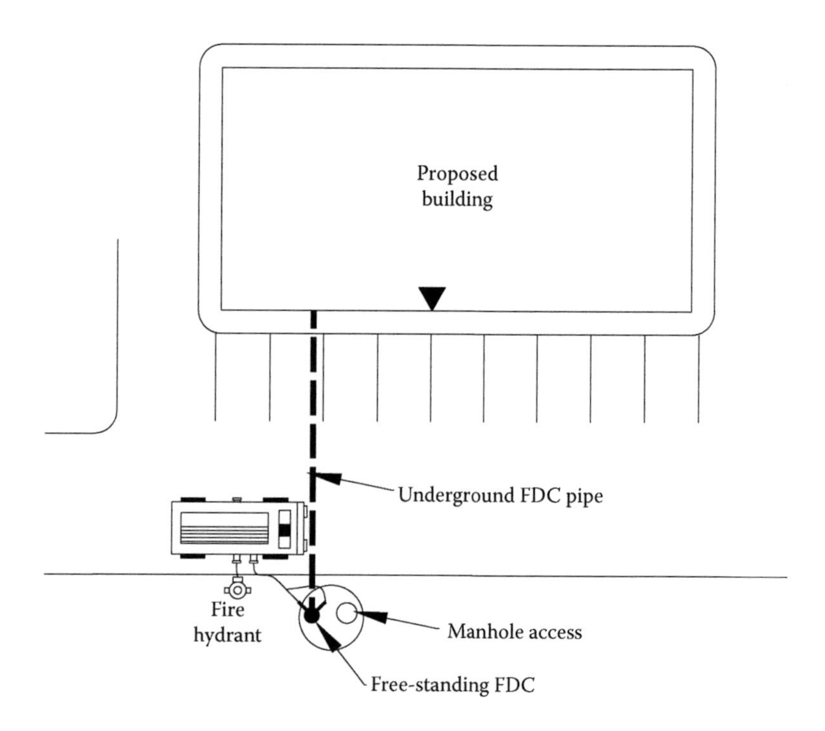

FIGURE 6.11 Free-standing FDC at fire hydrant.

(NFPA, 2011). For high-rise buildings, NFPA 14 requires two FDCs for each zone, unless approved by the AHJ. You may be thinking what is this zone, and what is the purpose of having two FDCs. Any building that requires standpipes (see your adopted building code for when standpipes are required) is considered to have at least one standpipe zone. The term *zone* here is used to identify pressure zones. To meet the high-pressure requirements to satisfy the standpipe demand, reduce cost associated with high-pressure pipe/fittings, and maintenance of pressure regulating devices, more than one standpipe pressure zone can be created (see NFPA 14 for more details).

A fire in a high-rise building is a major threat due to the unique challenges, such as mixed use, height challenges for aerial access, equipment transportation, high occupant load, longer occupant evacuation times, etc. Experts realize this and, in order to increase reliability, a secondary FDC connection is introduced. The primary FDC is located, as it should be, on the street front side of the building, while the secondary FDC is located remotely, on the perimeter, on another side. The purpose of this is that in case the primary FDC cannot be accessed, for example, due to falling debris and glass from the floor above, the secondary FDC is available. Both FDCs are interconnected.

CASE STUDY

Multiple apartment buildings, fully sprinklered, are proposed to be built off a public road, the Richard View Drive. The addresses of these buildings are #121, #123, and #125 Richard View Drive. The buildings are set back 60 feet from the road. A 20-feet-wide private road services the building rear, which also provides parking for the residents. Existing fire hydrants are to be utilized to serve the FDCs. Does Figure 6.12 appear to be correct?

The FDC locations are correct, as they are facing the address side. At first glance, the 100 feet distance is also met from the fire hydrant to the FDCs, but note the fence

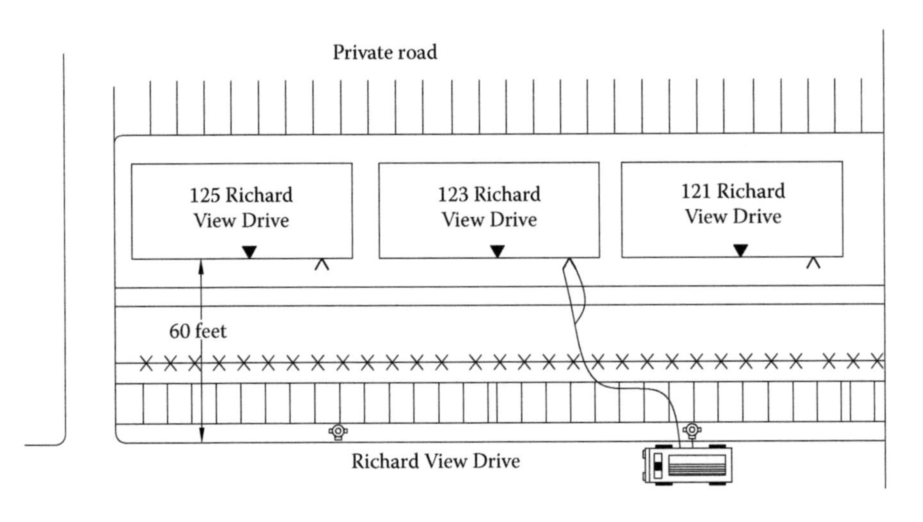

FIGURE 6.12 Example of proposed development.

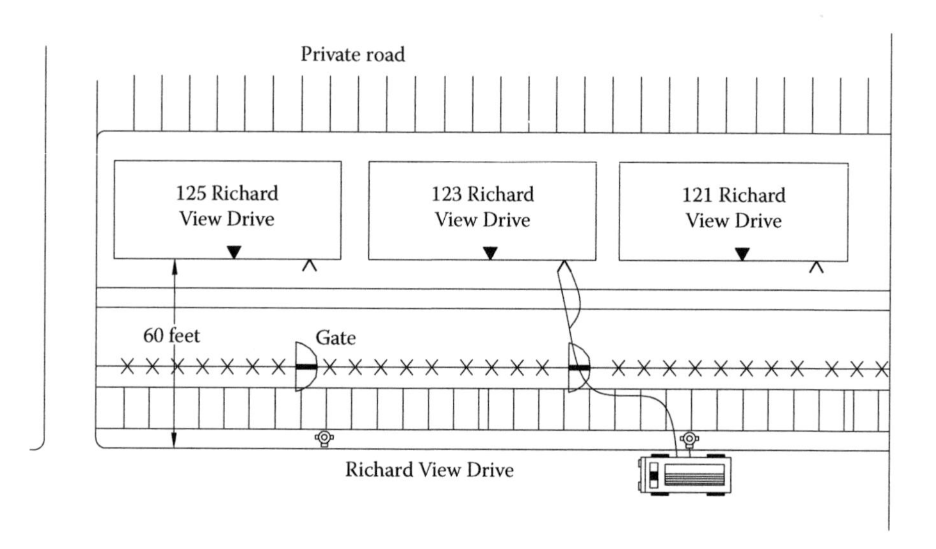

FIGURE 6.13 Access via fence.

(represented by the X marks). It is an obstruction. If this site was built as shown, the hose coverage of 100 feet would not be met, and at the time of an incident, significant delay would occur, allowing the fire to spread. Luckily, this was caught during plan review, and the engineer proposed an easy solution—to install gates that allow free egress. The hose reach can be measured across the gates to the fire hydrant, as shown in Figure 6.13.

On the basis of the above case study, it is vital for a plan reviewer to not only observe the information on the "Fire Lane Sheet" or "Fire Marshal notes," but also, as mentioned in Chapter 1, the site, geometry, layout, grading, and even landscape sheets which can contain information that greatly impacts the development.

Like fire hydrants, FDCs should also be protected using bollards if there is a chance of damage. Free-standing FDCs can be used if wall mounting is not feasible. Bollards must be designed to provide vehicle impact protection and must maintain at least 3 feet clearance from the free-standing FDC.

Not all buildings are required to have an FDC. Buildings that have a limited sprinkler system (not more than 20 sprinkler heads) can be connected directly to the domestic water system. Thus, for this case, an FDC is not required, as it can contaminate the water supply. Additionally, NFPA 13 states that buildings located in remote areas, inaccessible to the fire department, or where large-capacity deluge systems exist beyond the pumping capacity of the fire apparatus, or one-story buildings of 2,000 feet^2 or less, need not have FDCs.

PUMPER TEST HEADER

Pump test headers are utilized for mandatory testing (for fire pumps) where an on-site fire pump is present to boost the pressure for the building fire suppression system.

The pump test header connection can appear to be the same as an FDC; therefore, it is critical that the test header be located at a place away from the FDC, and that both the FDC and test headers are labeled correctly. In the past, numerous incidents have occurred where rookie firefighters caught up in the moment have mistakenly connected the pumper hose to the pump test header instead of the FDC. Pumping into a test header achieves no pressure boost due to the required shutoff valve, typically located inside the building on the test header line. Both the test header and the FDC must be shown and labeled on the site plan. The reviewer must assess the location and relocate as necessary to limit confusion.

FREE-STANDING FDC

The purpose of free-standing FDCs is to allow the FDC to be located at an accessible location (Figure 6.14). This can help locate it in a place where it is visible to the responders, away from obstructions, or near fire hydrants. Also, pump test header locations should also be kept away from exits, as they can create hazardous conditions when water discharges. In both cases, an underground pipe is utilized to make the connection from the building to the free-standing location. Profiles of this underground pipe should be shown and called out clearly on the plan. In areas subjected to freezing, a means to drain the system must be provided. See Figure A.6 in the Appendix for a detailed FDC connection in a pit, which should be provided with a site plan on a detail page for free-standing FDCs. In lieu of this option, one can also

FIGURE 6.14 Free-standing FDC.

FIGURE 6.15 Construction of free-standing FDC pit.

slope the underground pipe back to the building to drain to a lower level. Figure 6.15 is an image of a free-standing FDC and pit during construction (similar to the detail in Figure A.6 in the Appendix).

REFERENCES

NFPA. 2011. *NFPA 5000: Building Construction and Safety Code 2012 ed.* Quincy, MA: National Fire Protection Association, Inc.

Occupational Safety and Health Administration OSHA. 2006. *Fire Service Features of Buildings and Fire Protection Systems.* U.S. Department of Labor, OSHA 3256-07N, Retrieved from https://www.osha.gov/Publications/fire_features3256.pdf

7 Fire Truck Access

Every minute counts, whether responding to a fire or rescuing someone. Having quick access to the site and the building can make a big difference in response time. Shackelford, in his book titled *Fire Behavior and Combustion Processes*, indicates that it can be difficult and time consuming for the fire apparatus to access the structure not only because of fences, trees, and electrical wires, but also when buildings are set far back from the street (Shackelford, 2009). Consider the following: Imagine getting a fire call from a church on Christmas day. The first responders leave the station enroute to the site. The building is located far back in the property off the main road. Before the first unit enters the site, they see several cars parked on the street. Turning into the site, many cars are lined on the sides along the curb, reducing the width of the private access road to the building. The sheer size of the truck makes maneuvering a challenge. Upon getting to the site, the fire hydrant is blocked by a car. As other fire apparatus follows, they are unable to get to the rear from outside as they cannot get past the initial truck because of a choke point. The road is not wide enough because of the cars lined up on both sides. Firefighters have to manually carry heavy hose lines and equipment on foot to the area of the fire. The outcome: Every step has resulted in a significant delay, which has allowed the fire to grow and spread, consuming the entire building.

According to the U.S. Fire Administration, there are two parties involved during site development. On the one side are owners and developers, who want to maximize the use of their property and "minimize the impact and cost of drivable surfaces" (U.S. Fire Administration, 2009). Along with them are the architects/planners/engineers who design the site to make it most efficient and often esthetically appealing. On the other side, there are the code regulators and fire and safety personnel, who are looking for fast response access to protect the property.

The IFC defines a fire apparatus access road as a "road that provides fire apparatus access from a fire station to a facility, building or portion thereof. This is a general term inclusive of all other terms such as fire lanes, public and private street, parking lot lane and access way" (ICC, 2009). Therefore, any and all roads can be used as access, although certain characteristics must be met to qualify as an appropriate road to provide fire department vehicle access; these include the following:

1. At a minimum, the access road should be wide enough to provide access for the largest truck. The IFC, the Uniform Fire Code, and the NFPA 1 all require an unobstructed 20 feet width to allow for passage between parked trucks and engines, and where the pumper can hook up to fire hydrants to achieve proper hose connection without any bends or kinks. This width must be followed through all along the access road to all portions where access is required. Larger widths will be required when special apparatus is used, such as an aerial ladder truck. (Figure 7.1).

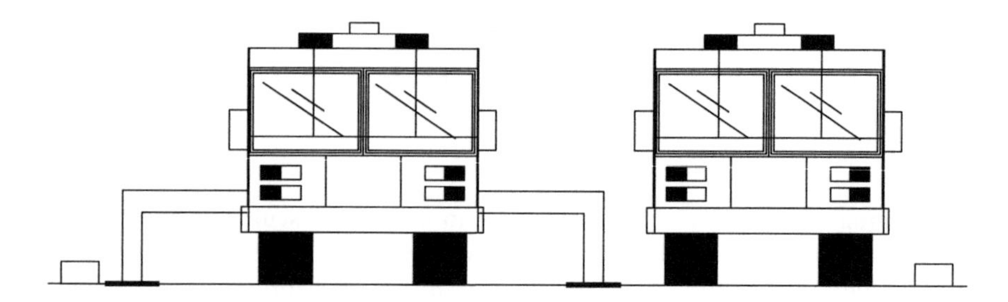

FIGURE 7.1 Engine crossing aerial ladder truck.

For areas that provide a setup for aerial ladder trucks, and where fire hydrants are present, a 26 feet width should be provided to allow passage for through traffic. Though aerial ladder trucks are only 8 feet wide, with the jacks spread, they can be 18 feet wide. See Chapter 8 for more information regarding aerial ladder trucks. Additionally, working room must be provided at the sides for personnel movement. There can be times when such widths are just not achievable due to site constraints, especially with existing buildings on site. In this case, mountable curbs can be used; for example, sidewalks can be used to achieve the lane width (Figure 7.2).

Another aspect to consider in the design of roads is the turning radius for the trucks. Make note of the inner and outer radii for the largest trucks you have in your jurisdiction (Figure 7.3).

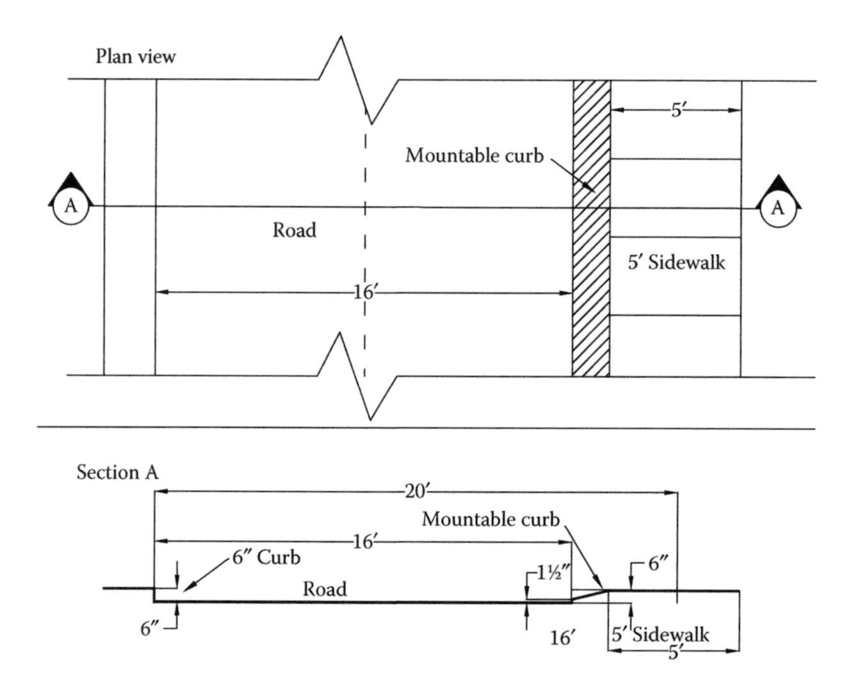

FIGURE 7.2 Mountable curb detail.

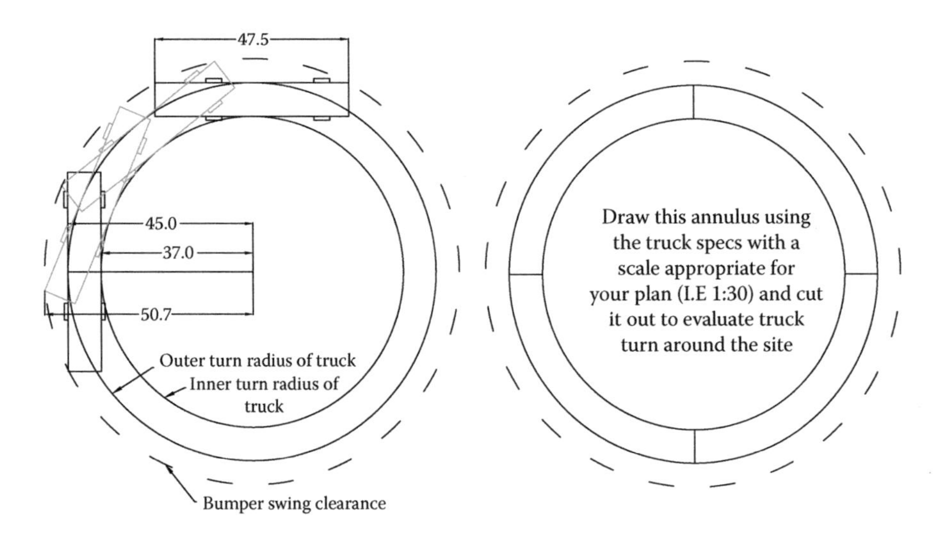

FIGURE 7.3 Truck turning dimensions.

No obstruction, such as fire hydrants, light poles, benches, and shrubs, shall be located in the bumper clearance. This is to prevent collision and damage, although curbs do not have to be considered, as they are generally lower than the bumper height (AlSaeefan, 2014). In new developments, the reviewer must pay careful attention to roundabouts to confirm that they allow for truck maneuverability.

2. Both the IFC and the Uniform Fire Code require roads to provide access to within 150 feet of all portions of the exterior wall of the story at grade, provided that the walls have openings such as windows and doors. *NFPA 1, Fire Code 2012* differs from this in that it requires a fire department access road to extend to within 50 feet of at least one exterior door that can be opened from the outside and that provides access to the interior of the building (NFPA, 2009). There are several reasons for this distance requirement.
   a. It allows firefighters to get in close proximity to the entry points to the building, such as rear and side exits, decreasing response time. Rescue efforts from other sides can be achieved if the main entrance is blocked. Furthermore, firefighters can get close to the area of the fire from outside to break windows to vent the fire and rescue trapped people in that area. There is also the advantage of using trucks to carry heavy equipment most of the way so that the personnel have to do less on foot.
   b. There is a limitation to the hose carried on the trucks. The closer they get, the more hose they have to use inside the building and the less the friction loss through the hose.
   c. According to the IFC commentary, 150 feet is the standard length of preconnected hoses carried by the fire apparatus. Temporary structures, such as tents, should be subject to the same fire access requirements.

Reviewers should also be aware of warehouses storing high-piled combustible materials. Depending on the area of the high-piled combustible storage, fire access doors are required along the exterior wall facing the fire apparatus access road according to IBC and IFC requirements, with the distance between them not exceeding 100 linear feet (ICC, 2009). Where the required distance is not achievable due to the terrain, topography, steep grades, etc., the fire official has the authority to consider an increase. Furthermore, if the building has an automatic fire sprinkler system, the sprinklers will be able to limit the spread of fire, so a larger traveling distance can be considered. This requirement can be considered as a code modification (see Chapter 9 for more information).

3. Exit doors and gates that cannot be opened from outside do not allow firefighters to gain entry and can severely delay response. When possible, a Fire access key box should be made available close to exterior doors and gates to provide entry. Security gates along fire department access must have a means approved by AHJ to provide free access. The IFC outlines details for gate access requirements. There are instances where gates are manned 24 hours. Despite this, it is better to have rapid entry boxes, since the guards can be occupied by the emergency, leaving the gates unattended and thus inaccessible.
4. On average, the height of the fire apparatus is 11 feet, but it can be higher. Both *NFPA 1, Fire Code* and the IFC require an unobstructed vertical clearance of not less than 13 feet 6 inches along the access road. The reviewer must examine any under passage, canopies, decks, balconies, and overhead power lines in and along the fire department vehicle access way. Where this distance is not achievable, reduced clearances can be approved with proper notification in the form of signs and headache bars.
5. Fire department access roads should be all-weather driving surfaces, designed and maintained to support truck loads. Often, owing to esthetics and street scape, grass pavers/grasscrete meeting the requirements for supporting the trucks can be used.

All access roads, including bridges that are used for fire access, must support the weight of the truck. The IFC appendix states that a weight of at least 75,000 lbs must be supported. Check the specifications for the largest truck you have in your jurisdiction and your neighboring response units.

Figure 7.4 shows a development of a new house deep within a property. The access road to the house is not only narrow, but also there appears to be a small bridge over a creek to get to the property. The reviewer must question whether or not the bridge can support the weight of the fire truck.

Bridges or decks of parking garages that are used for fire truck access must be able to support the weight. As a reviewer, or even an engineer, it may not be in your area of expertise to verify structural calculations. This is best left to a professional structural engineer. Simply make sure that a note is added to the fire lane plan to confirm that the truck weight is met. The reviewer must question areas of fire lanes with underground parking garages and utility vaults. These surfaces must be designed to

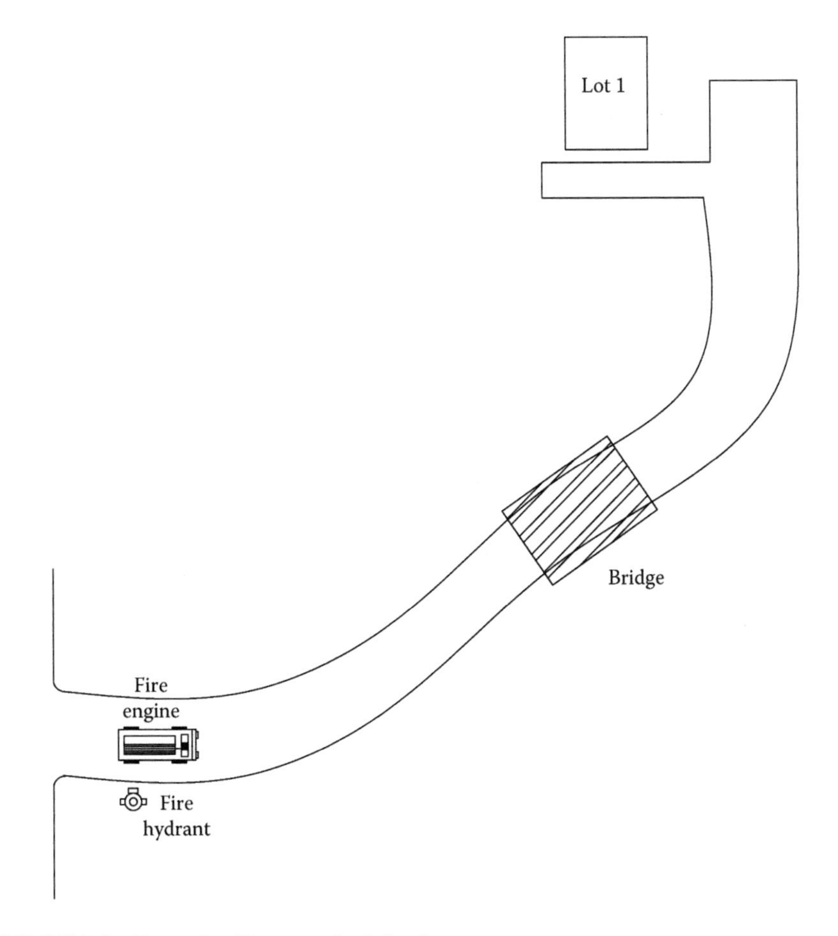

FIGURE 7.4 Example of house setback far from street.

support the weight of the fire trucks. If not, the fire access roads, especially those for aerial ladder setup, must be revised.

Lastly, it will be unwise to have a trucks setup right in front of the main entrance (especially aerial ladder trucks—Chapter 8) for buildings with large occupant loads, such as assembly occupancies. This is because large crowds may be evacuating via the entrance, and their path must be kept clear and unobstructed.

DEAD ENDS

Scenario—A fire is reported for a single-family dwelling unit. The building is located on a 20-feet-wide street along a pipe stem 200 feet off the main road (Figure 7.5). The units respond and put out the fire. Now it is time for the units to leave. As the driver is about to do so, he realizes that the road is a dead end and there is nowhere for him to physically turn the truck around. He is left to back up the truck for a long stretch without clear visibility.

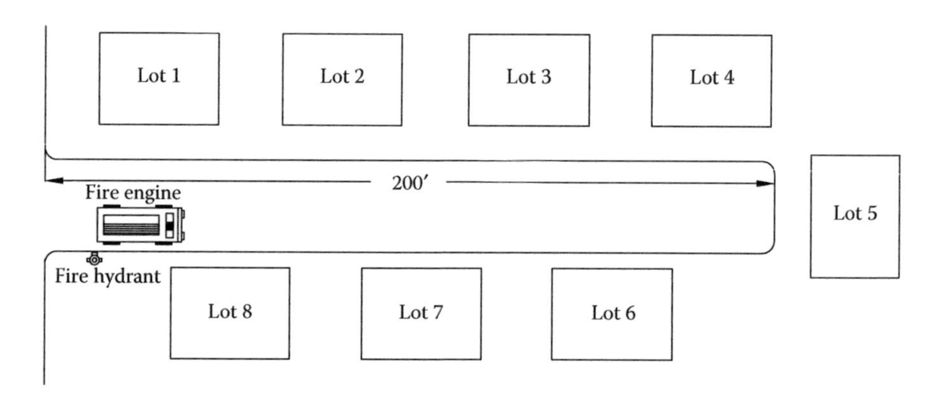

FIGURE 7.5 Long dead-end fire access road.

Consider the following cases of fire access and dead end roads:

1. Phoenix, May 2013: Firefighter Bradley Harper died as a result of being crushed between a fire engine and an ambulance. The engine was backing up, reversing down the road, and did not see the firefighter. Lack of a backup spotter to guide the truck was the cause. Had sufficient turnaround been provided, there would have been no need for the spotter, or for the truck to back up (Rossi, 2013).
2. Illinois, December 2012: A firefighter was killed at night by a fire truck backing from a narrow, steep inclined road to a rural brush fire that had extended to a vacant clubhouse. The "streets within the community were narrow, winding, and poorly illuminated at night" (NOISH, 2013). The victim had gotten off the truck to guide it backing up over a distance of 200 feet over the inclined road when he was struck and killed (NIOSH, 2013).
3. California, August 2004: After responding to a residential fire, a 25-year-old female firefighter died when she fell off the tailboard of the engine backing up. The engine was "backing to an intersection approximately 300 feet away in order to proceed forward" (NIOSH, 2005). The firefighter was helping with the backup operations to get the truck out of a dead-end site (NIOSH, 2005).

NFPA 1 states that "dead-end fire department access roads in excess of 150 feet in length shall be provided with approved provisions for the fire apparatus to turn around" (NFPA, 2009). There are many acceptable means of turnaround. One must consider the size of the truck when implementing the area of truck turnaround, the vehicle's length, inner and outer turning radii, and width (see Figures A.1 through A.3 in the Appendix and the IFC appendix). Turnarounds should be marked with no parking but should not have to meet the strict lane width requirement, as it will only be used for a single truck at a time. Using the annulus (see Figure 7.3), one can easily

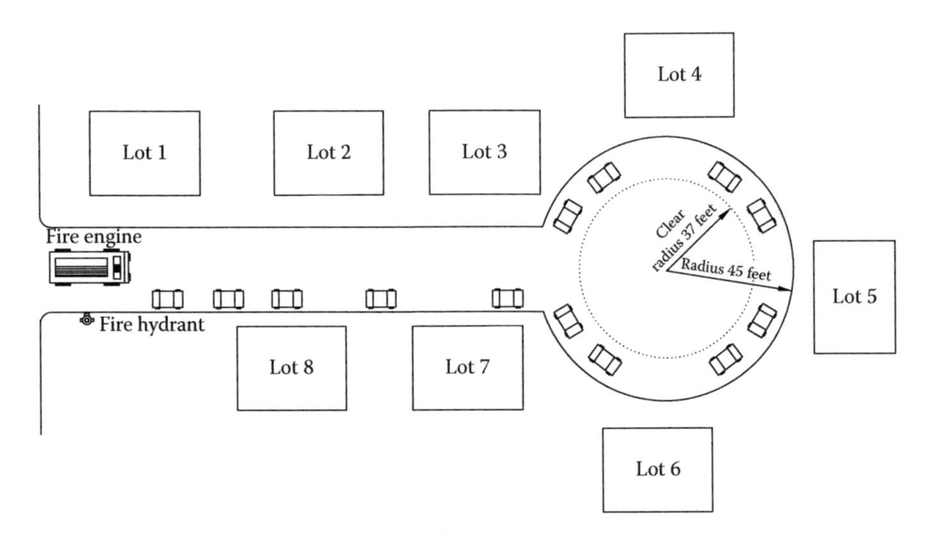

FIGURE 7.6 Cars blocking cul-de-sac.

check the turnaround. A cul-de-sac can also be used to provide adequate turnaround. One thing to keep in mind for any turnaround is that it should not be obstructed at any time. Cars parked in the cul-de-sac or in a hammerhead turnaround can render it useless. In Figure 7.6, with the parked cars, the fire truck cannot make an adequate turnaround due to the reduced radius of the cul-de-sac. It will have to back up along the long stretch of road.

Fire lanes can be used for through traffic and can be public or private roads, or even a parking lot. Any road that is not a public street and provides access for fire vehicles must be a fire lane and designed as such (i.e., all weather surface, weight capability, signage, and painting where needed to prohibit parking).

So what role does fire access play for the building that actually helps the building owner and developers? According to the IBC, when sufficient open frontage is met on a public way or open space of minimum 20 feet width, then the building is permitted an increase in area (basically allowed to be larger in square feet). Recall that the width required for fire lanes is 20 feet, so the minimum width of frontage is 20 feet. Additional increase occurs when the width is increased to 30 feet, while unlimited area can be established with additional increases (see your building code for conditions). A frontage increase occurs when more than 25% of the building perimeter is made accessible for the fire service. For the open space side to qualify, it must be on the same lot and must provide access via a private road, parking lot, or fire lane to within 150 feet of the perimeter wall, if not off a public way. Care must be taken when reviewing site plans with significant changes, such as deletion of roads and removal of open space via additions to the building. The area of the existing building on site should be reevaluated and the fire access that was provided before must not be diminished. According to OSHA, "Changing the amount of perimeter access can result in noncompliant building size" (OSHA, 2006), and thereby a noncompliant building.

FIRE LANE SIGNAGE

Any private road that provides access for fire apparatus should be marked with signs and yellow paint to designate a fire lane so that parking can be prohibited (Figures 7.7 and 7.8).

FIGURE 7.7 Utility trucks reducing access road width in a residential community. Trucks are parked in fire lane where signs are posted. (Courtesy of Fairfax County. With permission.)

FIGURE 7.8 Utility truck in fire access. (Courtesy of Fairfax County. With permission.)

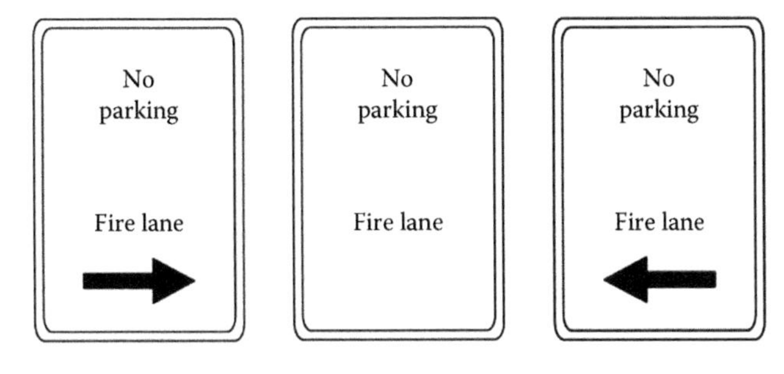

FIGURE 7.9 Fire lane signage.

Signs should be indicated on the fire lane sheet, with notes regarding painting on curbs. The reviewer must ensure that there is no parking within the designated fire lanes (fire hydrant locations must also be marked—see Chapter 4). Codes do allow for certain exceptions where the streets are wide enough to accommodate parking and truck passage. Typically, there are three types of signs as shown in Figure 7.9.

On the plan, the signs should be marked appropriately with proper designation along the fire lane and islands, with "arrow" signs at the beginning and the end of the fire lane, and "no arrow" in the middle. It is good practice to space out signs 70 feet apart along the long run of a fire lane.

It is incumbent upon the reviewer and the designers to find out which codes are adopted by the jurisdiction for access requirements, as significant changes to the site may be required. Another document to consult is the *NFPA 1141: Standard for Fire Protection Infrastructure for Land Development in Wildland, Rural, and Suburban Areas* for guidelines regarding such areas that differ from the abovementioned requirements.

REFERENCES

AlSaeefan, M. 2014. Engineering Solution for Fire Trucks. Saudi Aramco. Retrieved from http://www.nfpa.org/~/media/8CE7EBD32EE74ADF99A9392413F6F352.pdf

International Code Council (ICC). 2009. *International Fire Code 2009*. Country Club Hills, IL: International Code Council, Inc.

NFPA. 2009. *NFPA 1, Fire Code*. Quincy, MA: National Fire Protection Association, Inc.

NIOSH. 2005. Career Fire Fighter Dies after Falling from Tailboard and Being Backed Over by Engine—California. Centers for Disease Control and Prevention. Retrieved from http://www.lafire.com/lastalarm_file/2004-0814_Foster/2005-0520_NIOSH-Fatality Investigation/NIOSH_Fatality-Investigation.htm

NIOSH. 2013. Volunteer Fire Fighter Struck and Killed by Backing Fire Apparatus at Rural Brush and Structure Fire—Illinois. Centers for Disease Control and Prevention. Retrieved from http://www.cdc.gov/niosh/fire/reports/face201231.html

Occupational Safety and Health Administration (OSHA). 2006. Fire Service Features of Buildings and Fire Protection Systems. U.S. Department of Labor, OSHA 3256-07N. Retrieved from https://www.osha.gov/Publications/fire_features3256.pdf

Rossi, D. 2013. Lack of backup spotter blamed for death of crushed firefighter. CBS 5 (KPHO Broadcasting Corporation). Retrieved August 15, 2013 from http://www.kpho.com/story/22915731/lack-of-backup-spotter-blamed-for-death-of-crushed-firefighter

Shackelford, R. 2009. *Fire Behavior and Combustion Process.* Clifton Park, NY: Delmar Cengage Learning.

U.S. Fire Administration. 2009. *Coffee Break Training—Access and Water Supplies: Fire Apparatus Access Road Widths.* Retrieved November 23, 2014 from http://www.usfa.fema.gov/downloads/pdf/coffee-break/cb_fp_2009_1.pdf

8 Aerial Ladder Truck Access

Tall buildings create new challenges for fire operations. These include (Shackelford, 2009):

- Mixed use occupancies, which carry not only unknown fire loads but also people of different ages and with different physical challenges (i.e., the elderly, children, and handicapped people).
- Greater smoke movement in the building because of height and stratification.
- Transportation for equipment to upper floors.
- Fire fighters using the same stairways as those used by occupants for exiting the building.

Building codes today confront some of these challenges by specifying longer duration of fire resistance ratings for construction elements, increased number and size of exits to sustain the higher occupant load capacity, smoke control systems, and even the provision of fire service elevators on backup generators that can be used by the firefighters to reach the area of hazard. So the question arises: Is there a need for additional measures to contribute to addressing the above challenges?

Consider the following data on high-rise fires: "In 2007–2011, an estimated 15,400 reported high-rise structure fires per year resulted in associated losses of 46 civilian deaths, 530 civilian injuries, and $219 million in direct property damage per year with apartments, hotels, offices, and facilities that care for the sick being the top four occupancies" (Hall, 2013). This data does not indicate whether these buildings had fire suppression sprinkler systems, or if special design features, such as a smoke management system, were provided, but it is nevertheless still a serious concern. High-rise buildings in the past were not required to have many of the fire protection features that current building codes require. Furthermore, a building is not required to be upgraded, to have fire protection features in line with today's code. This chapter explores the above question in detail by examining aerial ladder trucks, a powerful tool that firefighters have as a resource to aid in rescue and fire suppression. At this point, you may be asking: What are aerial ladder trucks, and how do they differ from any other truck?

Aerial ladder trucks are advanced fire apparatus that have a built-in ladder that can be extended to floors beyond the reach of ground ladders (Figures 8.1 and 8.2). The ladder is used to transport firefighters and support equipment directly to (or close to) the area of the fire without having to go through the building, thus decreasing response time. Using the ladder, the firefighters can get themselves close to windows, breaking them if needed to vent heat and smoke. When the roof can be reached, they can also strategically cut holes to clear smoke/toxic gases from inside the building;

FIGURE 8.1 Aerial ladder truck setup and rescue. (From OSHA: Fire Service Features of Buildings and Fire Protection Systems. www.osha.gov)

they can use the ladder as a means of escape when trapped, and can even pull a hose line right up the ladder for fast attack. The ladder can also be used by firefighters to help rescue people who are trapped and have no way of exiting from inside the building, for example, via balconies. Additionally, the ladder is mounted with a fixed water spray nozzle, which can discharge up to 1,000 gpm at 100 psi in order to produce a master stream to put out and prevent the spread of fire. Ladder truck streams are thrown by a nozzle located on top of the ladder, which helps in preventing the fire from reaching the upper floors and spreading across the exterior surface. The hose streams also help to provide exposure protection to adjacent structures by running water down the surface from a safe distance (Figure 8.3).

Aerial ladder trucks have the ability to carry specials tools needed for operations, including "forcible entry, cutting and extrication" (NFPA Fire Handbook, 12-268). "From 1997 to 1999, municipal fire departments reported 6,300 aerial or ladder apparatus in use in the United States" (NFPA FPH, 12-269).

Aerial ladder trucks come in various sizes. You must be aware of the size of the truck used in your jurisdiction. The ladder on these trucks can extend up to 95 feet, but

FIGURE 8.2 Example of ladder on truck projecting water on fire surface. (Courtesy of Fairfax County. With permission.)

some have been designed to reach 200 feet, with the capability to rotate 360° in the horizontal plane (also known as turntable ladders). The elevations to which the ladder can operate ranges from slightly below the horizontal plane to straight vertical—90°.

Before the ladder can be used, it must be stabilized. This is done by four jacks, two on each side of the truck, which allow the weight to be distributed. With the jacks spread, the truck may require a 20-feet-wide road. Consideration should be given to where the ladder truck sets up, as wider roads are required for other fire

FIGURE 8.3 Example of ladder carrying fire fighters to guide projection of water on fire surface.

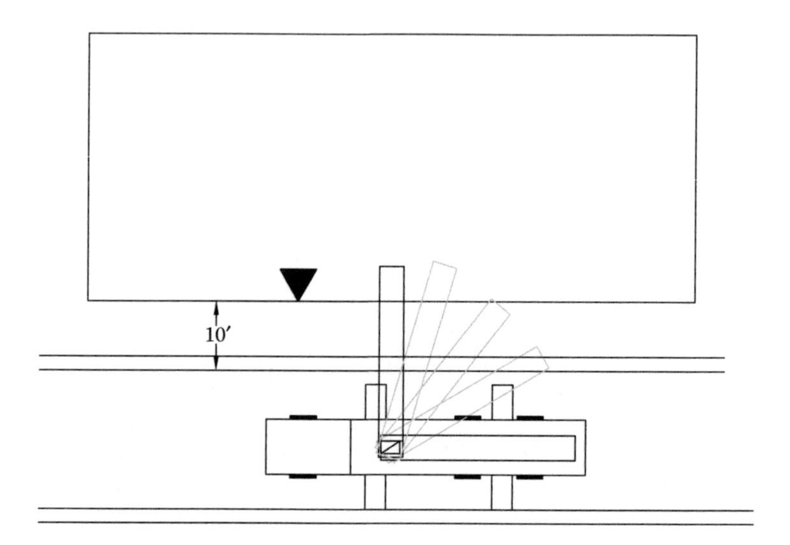

FIGURE 8.4 Plan view of ladder truck swing (verify ladder swing with your local aerial ladder specifications).

apparatus and passing traffic. The IFC appendix indicates a 26-feet-wide clear width for aerial ladder trucks. Furthermore, these trucks need to be set up on all-surface roads or grass-paving. Again, keep in mind the truck weight requirements and the weight distributed over the jacks during aerial operation. The jack stabilizers could punch through the surface if the surface is underdesigned, for example, over underground parking garages, which can cause the truck to tip over (Wieder and Smith, 2000). Owing to the angle required to swing the ladder, a minimum distance needs to be established between the truck and the building; being too close to the building will restrict the ladder's maneuverability. The IFC identifies that a minimum distance of 15 feet should be kept from the exterior wall to the edge of the truck access way (not the edge of the truck). Any closer than this and the ladder will have difficulty in achieving adequate swing (Figure 8.4).

The site of the building must be designed to allow for the largest trucks (turnaround and weight) as there can be a number of units responding to the scene, from different jurisdictions. Ladder truck access is not a code requirement, but it can be adopted via the fire code as an amendment to the IFC appendix by the AHJ.

So where will aerial ladder truck access be required? Buildings which ground ladders cannot reach should be considered as requiring aerial ladder truck access. Whereas it takes three firefighters to set up a 35-feet ground ladder, it takes only one to set up the aerial device (NFPA FPH, 2008), allowing for better allocation of resources. The IFC appendix states that buildings over three stories, or about 30 feet in height, should have aerial ladder truck access. Ladder trucks can also be required at the scene for buildings less than three stories or 30 feet. For example, in a house fire, the aerial truck can be used for providing a master stream onto the roof. Figures 8.5 and 8.6 show a ladder stream being used to both put out the fire and transport firefighters on a four-story apartment building.

FIGURE 8.5 Projecting water on roof. (From OSHA: Fire Service Features of Buildings and Fire Protection Systems. www.osha.gov)

Where large open spaces and parking lots are not available, it is best to lay out aerial ladders to allow for longitudinal access, meaning that, rather than having the truck long side parallel, provide a nose-in access location to back into. This allows the truck more stability and maximum reach (Wieder and Smith, 2000). The site can be designed to provide nose-in access if the building is set back from the road (Figure 8.7).

The distance of the truck setup from the building is inversely proportional to the stories it can reach. Basically, this means that the further the truck is away from the

FIGURE 8.6 Apartment rescue.

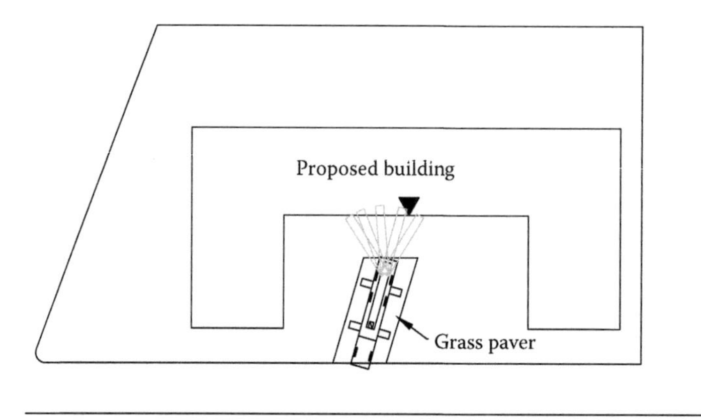

FIGURE 8.7 Nose-in access in the form of grass pavers can provide access.

building, the fewer number of stories and area it will cover. The IFC recommends that there should be a maximum distance of 30 feet from the outside of the truck access lane to the face of the exterior wall (the exterior wall here is the wall for the portion of the building above 30 feet or three stories where aerial ladder access is required—not the lower stories). The truck can set up anywhere along the access lane width to achieve the best angle of swing and height.

The reviewer must evaluate the building for any changes in stories (see Figure 8.8). With the taller portion being in the middle, the lower stories prevent access to the upper levels. In this case, the building may need to be redesigned. Determining the access is very important, and should be factored into the conceptual design phase of the building so that changes can be made easily.

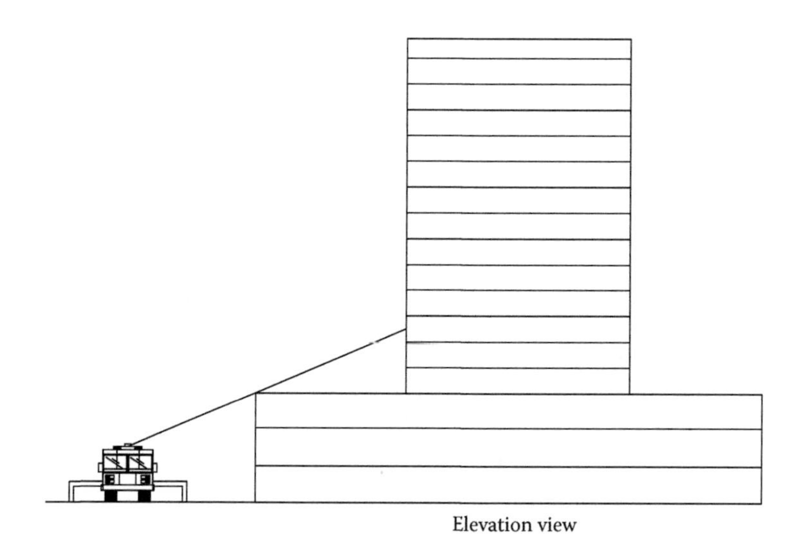

FIGURE 8.8 Example of building preventing ladder reach.

As shown in Figure 8.8, with the lower floors stepped out (also check for canopies), the ladder cannot reach levels four and five. Though the angle above may not be optimal for climbing firefighters, it can still project fire streams.

When analyzing aerial ladder truck access, it is important to be aware of the streetscape and the environment. A landscape or tree preservation plan will show trees, along with many other features of the site, that may not be found on the site plan or fire lane sheet. These can have a direct impact on fire access. Imagine having a row of large canopies that restrict aerial ladder truck setup, or water features, benches, or poles. Trucks should have access so that coverage can be maximized. The distance between the trees should be checked, with fully grown canopy sizes, for truck setup and ladder swing (Figure 8.9).

Note how tree height and placement affect access for the ladder to the story. Both short and tall trees placed in between the access lane and the building will prohibit the ladder from reaching floors four through seven, respectively, in Figure 8.9. Typically, trees less than 15 feet in height are not a problem.

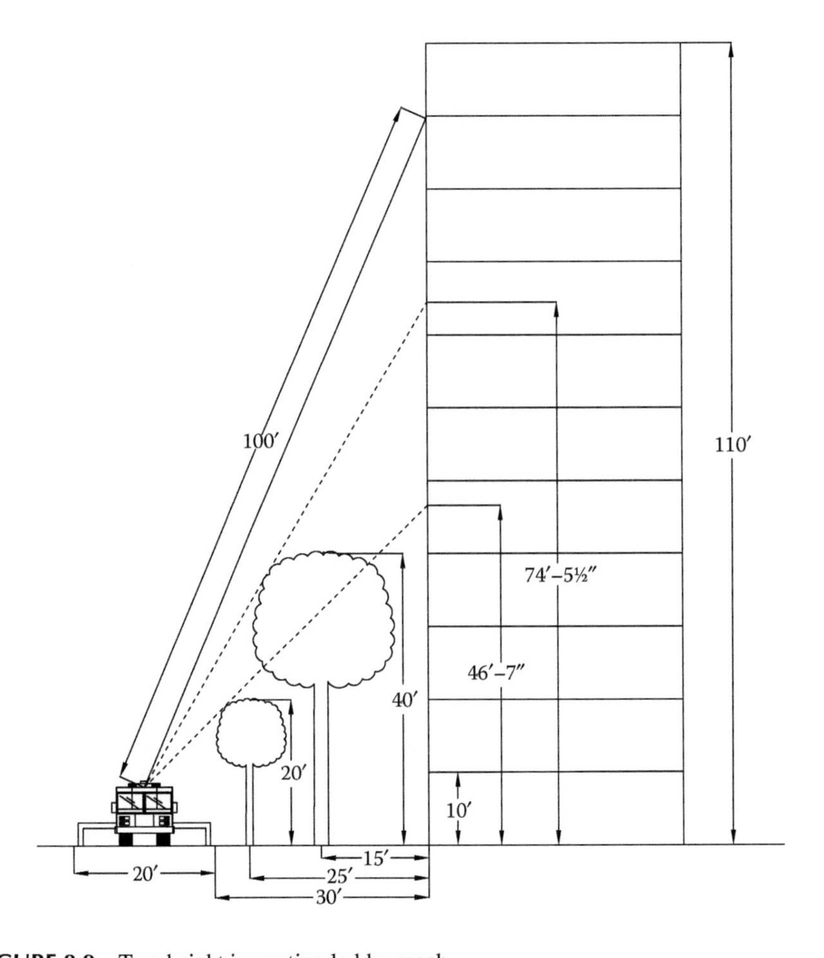

FIGURE 8.9 Tree height impacting ladder reach.

Even with the best positioning, the spread of trees can impact the perimeter and scrub surface of the building accessible by the ladder (Figure 8.10). Typically, the landscape plan shows fully grown tree canopies. Trees spaced closer together, and those with large canopies, prevent the truck from achieving the needed turntable

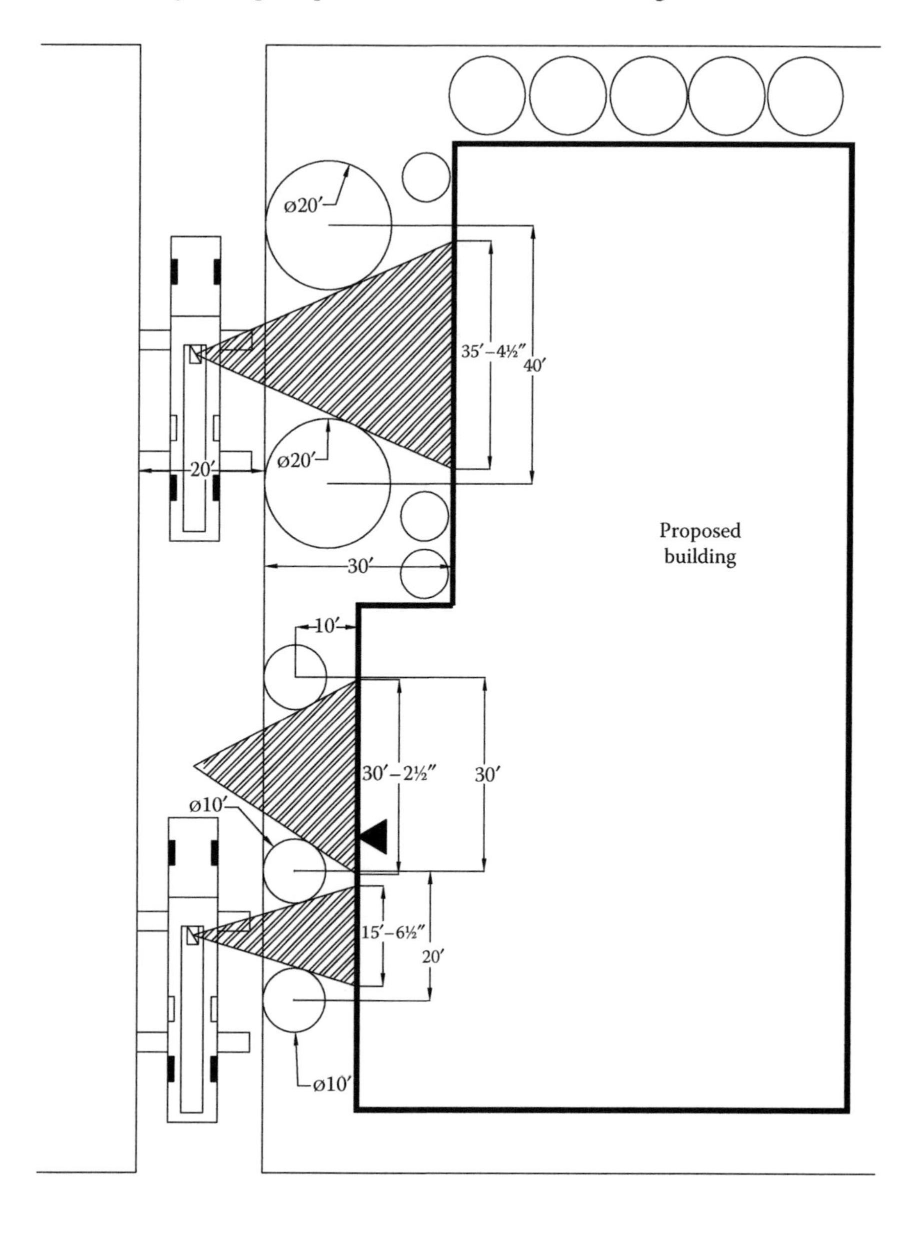

FIGURE 8.10 Access area in between trees. Shaded triangles represent accessible areas. Circles represent tree canopies and show their size.

space. As a reviewer, you must verify where access is to be provided for the aerial ladder and ensure that it is not compromised.

Can the fire personnel cut down trees when needed? Most likely they can, but here are two things to consider:

1. They are not expert tree cutters, and they have not had any training in cutting trees. A tree falling in the wrong direction can hinder fire operations, especially if power lines are nearby, and injure bystanders and people exiting the building. In March 2014, in North Carolina, an engine company was responding to a call; while en route, the road was obstructed by a falling tree. "Emergency personnel were cutting the tree from the vehicle when another tree fell on the truck" (Lopez, 2014). With power lines involved, two firefighters were trapped for over 45 minutes while other units responded and helped them get out safely.
2. Every second spent cutting the tree is a second that the fire grows, which could have been used to save someone's life and limit property damage.

With so many buildings going green and energy efficient, fire access must be carefully evaluated. Energy-efficient windows can be very difficult to break for venting operations. Buildings have significantly changed over the decades to allow for better insulation and to retain more heat, including "the installation of double-paned glass to better insulate the interior from both heat and cold" (Shackelford, 2009). Because of security, there can be buildings (e.g., government) that have blast-proof windows. Most of the time, these windows are installed from ground level up for a few stories only. Where possible, an identifier mark should be placed on these windows so that firefighters are aware of which can be broken. There is no point in providing ladder truck access to the side of the building where there are no windows, balcony openings, or access. The reviewer must pay close attention to this. When fire separation is not adequate, building codes require that walls be rated and protected (Figure 8.11).

Another thing to consider is the utility plan, which will show light pole locations and power lines. Areas where overhead power lines run should be located and not be considered for aerial ladder access.

HOW MUCH ACCESS IS ENOUGH ACCESS?

How much access should be provided for aerial ladder trucks, and what is considered access? We are out of the era where all the buildings were squares or rectangles with all-around access. There are many unique building designs, streetscapes, and zoning regulations that shape the building and the site. First of all, remember that aerial ladder trucks are for buildings or portions of buildings that are beyond the reach of ground ladders; this is the first step. After this, access should be looked at in two ways:

1. Access required by the truck to get around the building: For new developments, the IFC appendix recommends that aerial ladder truck access shall

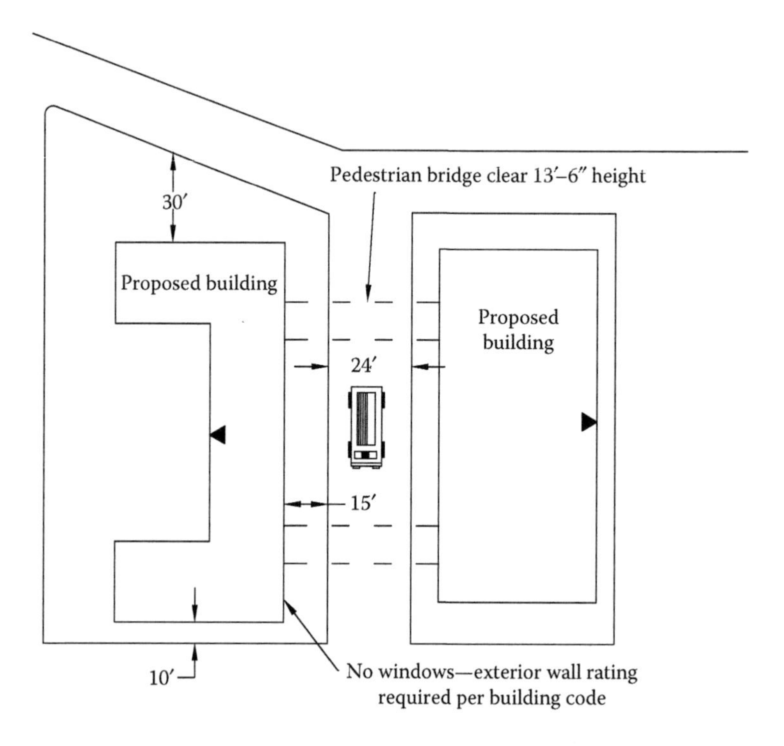

FIGURE 8.11 Confirm access is not given on sides where walls have no windows.

be provided parallel to at least one entire side of the building, although this must be examined further. There are several units and vehicles that respond to an emergency. This could consist of engines, pumpers, tankers, aerial ladder trucks, ambulances, etc. If fire department access is provided to only one area or side—for example, the main entrance—it will create a choke point, causing congestion, which will inhibit proper truck setup, obstruct the egress path for those exiting the building, and prevent the setup of adequate hose lines. Additionally, the secondary responding units will have to park further away. Remember, firefighters are carrying a lot of equipment and gear, including hoses, 2½-gallon extinguishers, forcible entry tools, radios, flashlights, and nozzles, so time is working against them; having them park further away will delay operations. Responders are trained to go to the front (Alpha side) and rear (Charlie side) of the building, usually where the main and rear exits are. Therefore, it is good practice to have aerial ladder vehicle access to at least two sides.

2. Access required for the ladder: Codes are silent on how much ladder coverage is required for buildings. Should it be the entire front or the entire perimeter of the building? The first thing a reviewer must do is to evaluate the type of construction, the use group(s), the height, the perimeter, and whether or not the building is sprinklered. In an ideal world, getting 100%

ladder access to the entire exterior surface will be best, as the fire can occur on any side, with occupants trapped on any side. Recall the example in Analysis section of the town house fire of a building under construction with no rear access. Furthermore, systems can fail and the fire can have a devastating impact. Despite all this, not every situation can be accounted for and designed into the site. Consider the case of a building over twenty stories; even at the best position, a 100 feet ladder cannot reach 12–20 stories (assuming approximately 10 feet 2 inches per story), as it is beyond the maximum reachable height. Additionally, not all sites can give 100% access due to terrain, site limitations, and zoning regulations.

The following are some practices to follow:

- Buildings that are not fully sprinklered should be provided with all-around access.
- Corner locations are preferably best as they allow access to at least two sides of the building (Wieder and Smith, 2000).
- Access points to windows, such as those serving large corridors, can help firefighters get in fast, and can be used to ventilate, to stage rescue, and to stretch hose lines.
- For wooden constructions, especially nontransient residential-type occupancies with balconies, access should be provided to all sides. This is because most fires occur in residential occupancies (cooking and smoking) and due to the ease of ignition of wood, and the high fuel load, there is a higher risk of persons being trapped and the fire spreading. In addition, hotel and apartment occupants have only one way out from the unit, which is via the corridor (for units with only a single exit).
- Building codes have introduced special provisions that allow a lot of flexibility in building design. Podium/pedestal buildings are such building, which can be designed for one building on top of another (basically a single structure consisting of more than one building). These buildings can be built with noncombustible, fire-resistive construction below, with a fire separation between a combustible wooden frame construction on top. Furthermore, if the building on top is less than 60 feet in height above grade, and not more than four stories (measured over the fire separation), the codes allow the sprinkler system to follow NFPA 13R, which omits sprinkler heads in spaces such as closets, bathrooms, and attics, where a fire can spread unnoticed (recall the example in Analysis section).
- Access should be provided along the length of the fire access roads serving the building, or to the maximum length possible, as falling debris and broken glass from the fire can prevent truck setup in a defined location.
- Commercial office buildings are at lower risk compared to residential occupancies. In residential occupancies, occupants may be sleeping, can be under the influence of alcohol or medication drugs, and have a tendency to cling to their belongings (Wootson, 2015).

Many people debate the need for aerial ladder trucks. Several arguments have been presented for new construction buildings, especially for high-rises. Some of these are as follows:

- The building has an automatic fire sprinkler system that controls the spread of fire.
- A smoke control system in place to prevent smoke egress into the path of travel allows for a tenable environment.
- A fire alarm to detect fire and voice evacuation to notify occupants with clear instruction allows for early fire response when the fire is at an incipient stage.
- Built in fire command center allow control over key buildings systems, that is, control of elevators, HVAC, and smoke pressurization system.
- The building is of fire-resistant, noncombustible construction.
- The rated exits and passages allow for safe and secure discharge.
- Emergency generators prevent power interruptions of life safety and fire protection systems.
- Separate to the main elevators, a fire service elevator allows for the transport of firefighters and tools quickly and easily.

Another issue is that ladder trucks cannot reach high floors beyond the reach of the ladder. So what purpose does aerial ladder access serve for high-rises?

All the points mentioned above are valid and every feature mentioned above does enhance the safety of the building. It is also correct that the ladders can only reach certain floors even in optimal setup locations, however they can help firefighters quickly get to higher floors through exterior means instead of going up the same routes where people are exiting. Besides this, the objective is safety, since lives are at stake. If there is a single chance that these ladder trucks can assist firefighters, then they should be provided. Additionally, if any of the above features fail, or there is an inoperable or ineffective sprinkler system (see Analysis section), then manual firefighter response and rescue is the only option available. Damaged underground fire lines can disable the sprinkler water supply without anyone noticing; smoke and toxic gases entering the stairways (due to doors left open by occupants) will expose both fire fighters and those trying to evacuate to toxic gases. Additionally, fire service elevators can only carry so many personnel in full gear and with equipment—that is, high-rise packs, oxygen tanks, etc.—and if the elevator becomes disabled (due to the fire spreading to electrical lines), then the only way to get up to the fire floors is via the stairwells. It is not an easy task to climb twenty-five stories, and not so easy to go down again to get equipment once up. Recall the One Meridian plaza fire in 1991, where three fire firefighters perished after climbing up twenty-eight stories, where they experienced fatigue, became disoriented from the smoke and toxic gases, and ran out of oxygen tanks (FEMA, 1991).

Consider the following cases:

Case 1: October 2014—a high-rise fire took place in downtown Detroit on the sixth floor of a twelve-story building. The floor was under reconstruction for a new tenant layout. A construction worker reported that the insulation had caught fire. The fire sprinkler system was turned off due to the work being done. Fire service personnel were not told of the sprinklers being shut off on that floor (Scripps Media TV Station Group, 2014).

Case 2: Dubai, UAE—In February 2015, the super-tall residential structure, Torch, in Marina, standing at approximately 1,022 feet (336 m), eighty-five stories, with 676 apartments, experienced a fire at 2 a.m.—thousands of occupants had to be evacuated after a fire broke out on the fiftieth floor. The cause of the fire appears to have been either a cigarette or a coal from a hookah. Owing to the construction type (noncombustible), it is most likely that sprinkler heads were omitted from the balconies. Either the external flammable cladding or air-conditioning vents appear to have spread the fire, along with high winds. Debris and glass fell down below. Fortunately, there were no deaths reported. A neighboring high-rise building was also evacuated as a precautionary step. The building was equipped with fire alarms, which alerted the occupants to egress quickly, although some occupants came out 4 hours later, after sleeping through the alarms. The lifts were not working, so occupants had to traverse down stairs, where they passed the response units going up. Occupants described smoke in stairwells and sprinkler heads discharging in the corridors (Aljazeera, 2015; Harriet, 2015).

Case 3: June 2008, Las Vegas, Nevada—A fire lasted for more than 1 hour before being suppressed in a thirty-two-story hotel containing 3,000 resort rooms. The building was a high-rise of noncombustible construction, built according to the 1991 Uniform Building Code. The fire appears to have started from a welding operation on the roof, where exterior cladding and decorative material caught fire. The fire spread rapidly up, down (from falling debris), and laterally over 80 feet, breaking windows from floor 32 down to floor 29. Interior sprinkler operation prevented the fire from spreading inside. Over 100 fire personnel responded. Luckily, there were only seventeen people injured and no fatalities. Manual firefighting from inside the building and from the roof put the fire out (Beitel and Evans, 2001; *USA Today*, 2008).

In light of the above factors, at a minimum, one access point (without any obstructions) should be provided to at least two sides of a high-rise building.

What if access cannot be met? There may be times when, owing to site constraints, access cannot be achieved, for example, for both the front and rear, or to within 15–30 feet. For this, a code modification (see Chapter 9) request should be sought. For example, extra stairs or elevators can be provided for fire service use only.

REFERENCES

Aljazeera. 2015. Thousands evacuated after fire engulfs Dubai tower. Retrieved March 2, 2015 from http://www.aljazeera.com/news/2015/02/thousands-evacuated-fire-engulfs-dubai-tower-150221012315584.html

Beitel, J., and Evans, D. 2001. The Monte Carlo exterior facade fire. *Fire Protection Engineering*. Retrieved March 20, 2014 from http://magazine.sfpe.org/fire-investigation/monte-carlo-exterior-facade-fire?utm_rid=CPG04000000872990&utm_campaign=1837&utm_medium=email

Harriet, A. 2015. Dubai Marina Torch Tower fire: Londoner tells of how he slept through a towering inferno. *The Telegraph.* Retrieved March 22, 2015 from http://www.telegraph.co.uk/news/worldnews/middleeast/dubai/11427715/Dubai-Marina-Torch-Tower-fire-Londoner-tells-of-how-he-slept-through-a-towering-inferno.html

FEMA. 1991. Highrise office building fire One Meridian Plaza Philadelphia, Pennsylvania. Retrieved March 21, 2015 from www.usfa.fema.gov/downloads/pdf/publications/tr-049.pdf

Hall, J. 2013. *High-Rise Building Fires.* Quincy, MA: National Fire Protection Association. Retrieved October 8, 2014 from http://www.nfpa.org/research/reports-and-statistics/fires-by-property-type/high-rise-building-fires

Lopez, R. 2014. Firefighters trapped by falling tree—Responding to tree on vehicle. News & Record (Greensboro, NC). *Firefighters Worldwide.* Retrieved November 4, 2014 from http://www.firefightersworldwide.com/2014/03/firefighter-tree/

Scripps Media TV Station Group. 2014. Detroit firefighters put out high rise fire at Detroit Savings Bank building downtown. *WXYZ Detroit.* Retrieved from http://www.wxyz.com/news/region/detroit/detroit-firefighters-battle-fire-at-detroit-savings-bank-building-downtown

Shackelford, R. 2009. *Fire Behavior and Combustion Process.* Clifton Park, NY: Delmar Cengage Learning.

USA Today. 2008. 17 hurt in fire at Vegas' Monte Carlo. Retrieved March 20, 2014 from http://usatoday30.usatoday.com/news/nation/2008-01-25-vegas-fire_N.htm

Wieder, M., and Smith, C. 2000. *Aerial Apparatus Driver/Operator Handbook.* 1 ed. IFSTA. Stillwater, OK: Fire Protection Publications.

Wootson, C. 2015. Charlotte high-rise growth creates firefighter challenges. *Charlotte Observer.* Retrieved from http://www.charlotteobserver.com/news/local/article 13133225.html

9 Code Modifications

There are times when you will find a situation that just does not meet the code. Perhaps it is something that could not be achieved due to design challenges, such as no fire flow availability on site due to there being no public water access, or possibly access restrictions around the building because of the topography. For new buildings, it is easier to make changes to design, but for existing sites, such modifications can be challenging. Here "code modification" comes into play. A code modification is not a waiver of the code requirement; rather, it is a requirement to satisfy equivalence, which allows for comparable change for the replacement of the item not meeting the code (a give and take). An example would in the case where fire access is not met for a building. If the building was not required by the code to have a sprinkler system, then it might be proposed to add a sprinkler system. So, as it will take the emergency unit extra time to get to the fire location, adding a sprinkler system will control the spread of the fire and thus allow the extra time necessary for firefighters to make it to the site. It is important to have proper documentation and approval for code modifications via the correct chain of command. The modification only applies to an individual case and cannot be carried over to the next project. As mentioned, each site and building is unique.

Example: An existing mechanical room serving the pond fountain has access from an existing gravel road (Figure 9.1). The gravel road is to be demolished for a new building. The mechanical room building does not have permanent occupancy, and the pump is shut off during the winter season. A new 20-feet-wide access lane cannot be provided to the left of the pond due to a steep hill. How can access be achieved so that the mechanical room building can meet the code?

Solution: The mechanical room building is not permanently occupied and is only seasonally used. One approach would be to provide an 8-feet-wide access lane to the left of the pond to accommodate firefighter foot access. Additionally, an underground horizontal standpipe should be provided. This standpipe will allow the fire pumper to charge the side nearest to the street via the pumper and a fire hydrant, so the firefighters can get water from the outlet FDC near the building, saving them the time of stretching long hoses (Figure 9.2).

Example: A site is proposed for a residential building with a parking garage (Figure 9.3). The parking garage and the residential building are separate buildings, as per the building code, with different construction types. Fire walls are established between the garage and the residential building. The garage is proposed to be open, and meets the requirements of the building code, based on openings and ventilation. Since the garage is open, it is not required to have a sprinkler system but does have a manual dry standpipe. The residential building is fully sprinklered. Is access met?

Since these are separate buildings, they must be looked at individually. Access is met for the residential building, but not for the garage. With the garage wrapped by

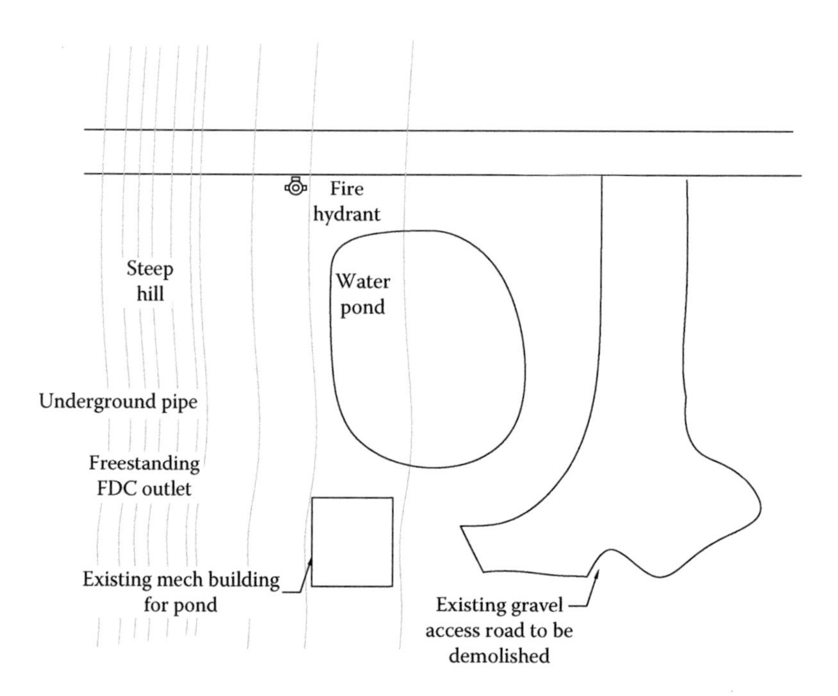

FIGURE 9.1 Site layout showing existing conditions.

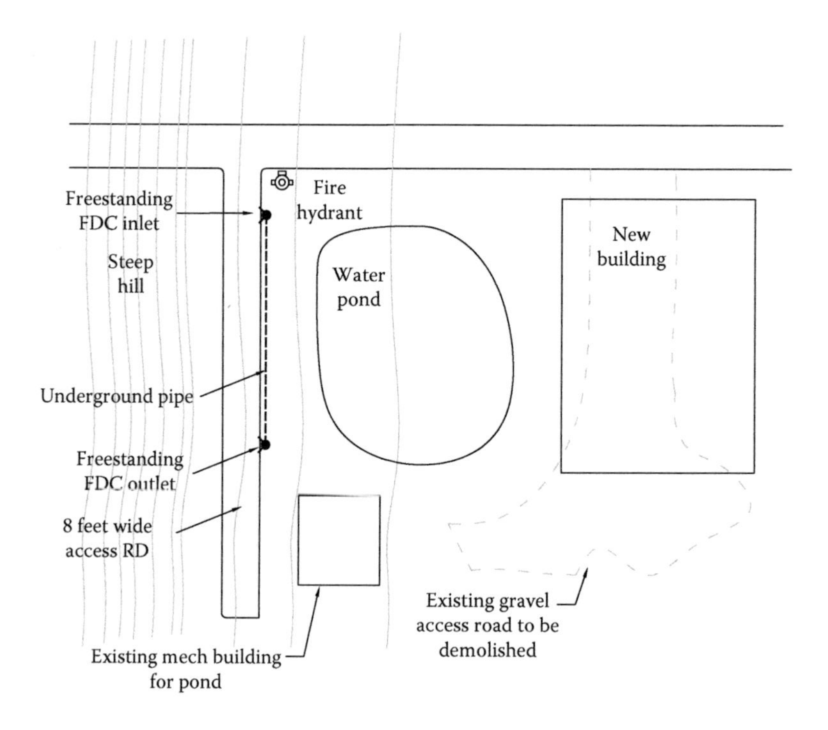

FIGURE 9.2 Site layout showing proposed solution.

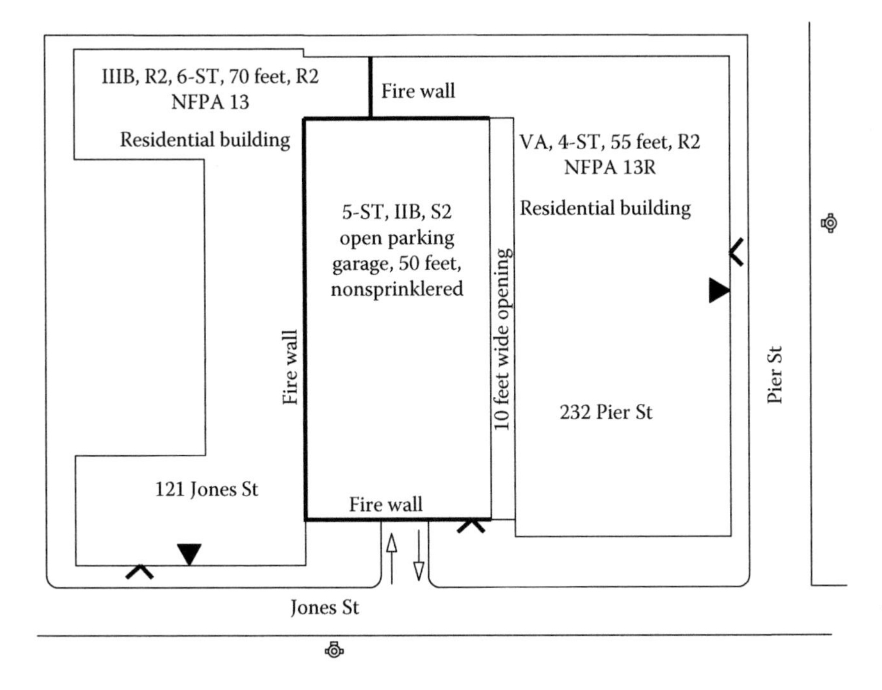

FIGURE 9.3 Site layout showing a proposed building.

the residential building, there is no truck access or aerial ladder truck access. The garage is not in compliance with the code. A code modification would be to sprinkler the parking garage with a dry sprinkler system. This will help contain the fire, giving firefighters more time to reach it.

10 Conclusion

How important is a site plan review from the fire protection point of view? It can make the difference between saving someone's life and property and not doing so. Ensuring that a site has adequate access to the fire department, has accessible and properly located FDCs, has properly positioned fire hydrants, has positions for ladder trucks, and has full provision of fire flow water on site—all these factors are critical to ensuring that emergency personnel can do their job correctly and rapidly. A fire site plan review can be summed up as follows. For owners, it will account for the safety of the occupants in their building and for the protection of the building; for architects and engineers, it will lead to a cost-effective code compliance design; and for plan examiners and inspectors, it will be a tool to tactically evaluate the site for optimal fire response and emergency personnel. In this book, I have tried to capture the important aspects and best practices that should be reviewed and designed into the site for fire protection.

Appendix

FIRE TRUCK TURNAROUND EXAMPLES

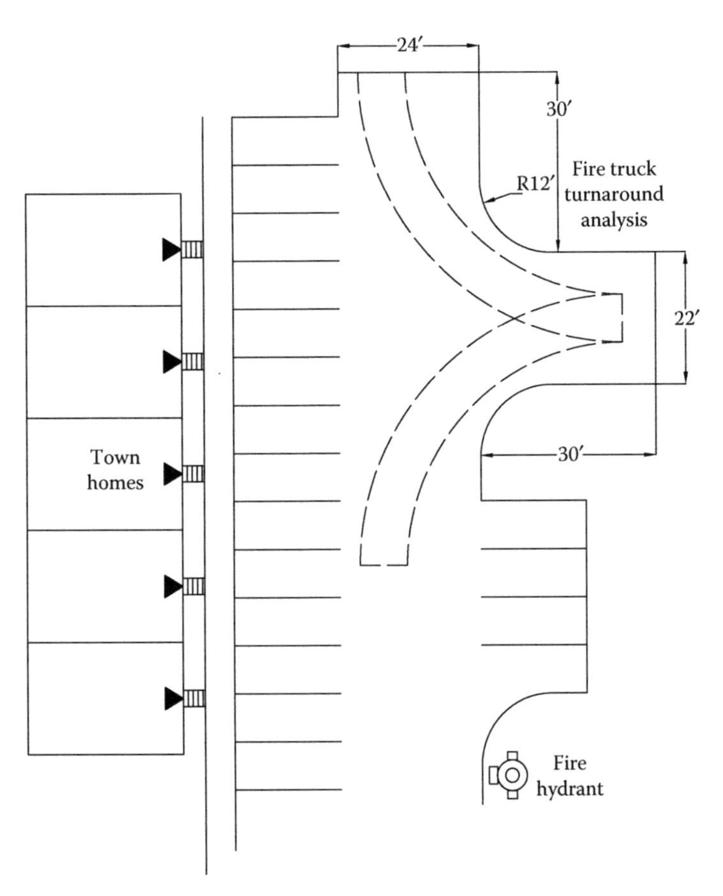

FIGURE A.1 90° reverse fire truck turnaround.

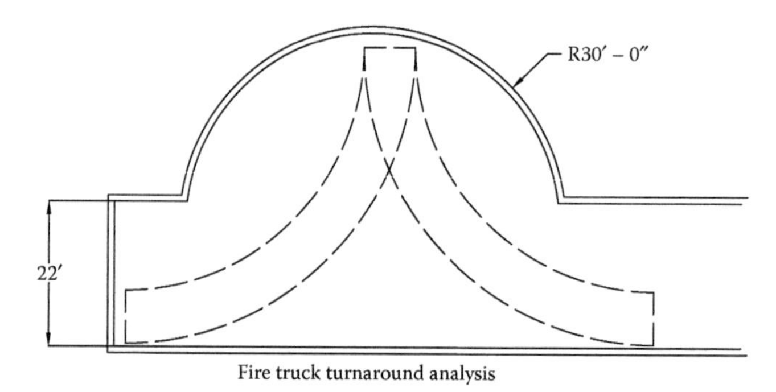

FIGURE A.2 Reduced radius—adequate turnaround is met.

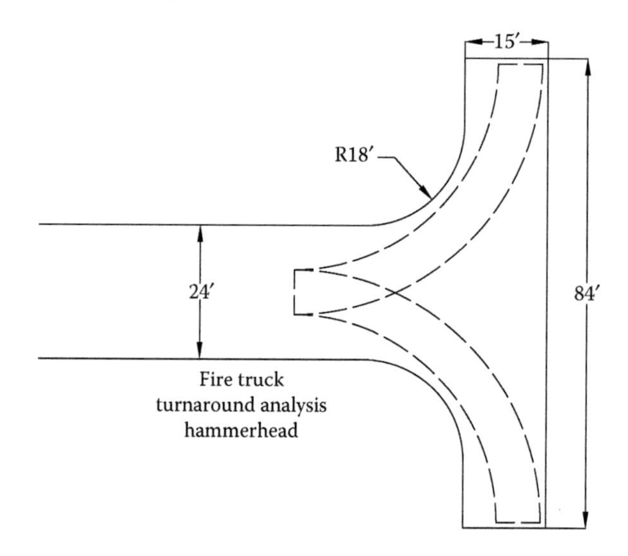

FIGURE A.3 Acceptable method of truck turnaround in hammerhead.

FIRE HYDRANT DISTANCE FROM CURB

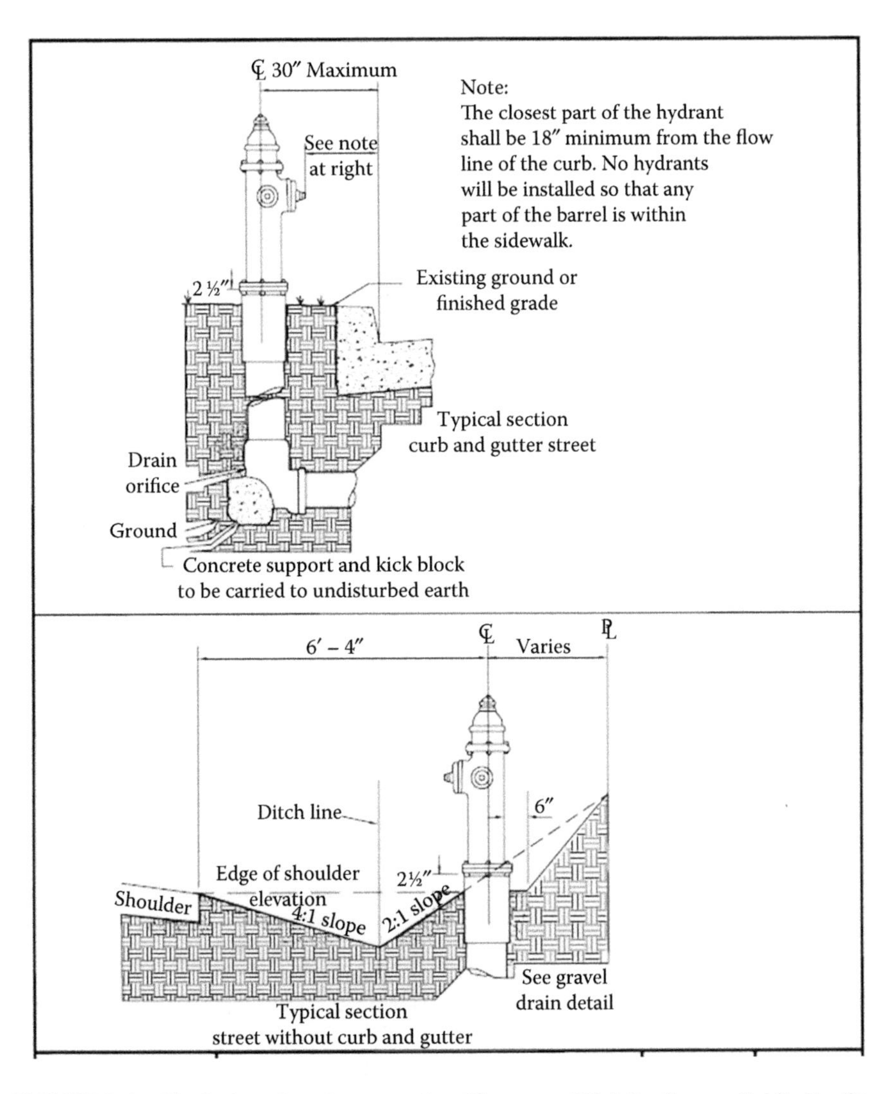

FIGURE A.4 Fire hydrant location examples. (Courtesy of *Fairfax County Public Facilties Manual*. With permission.)

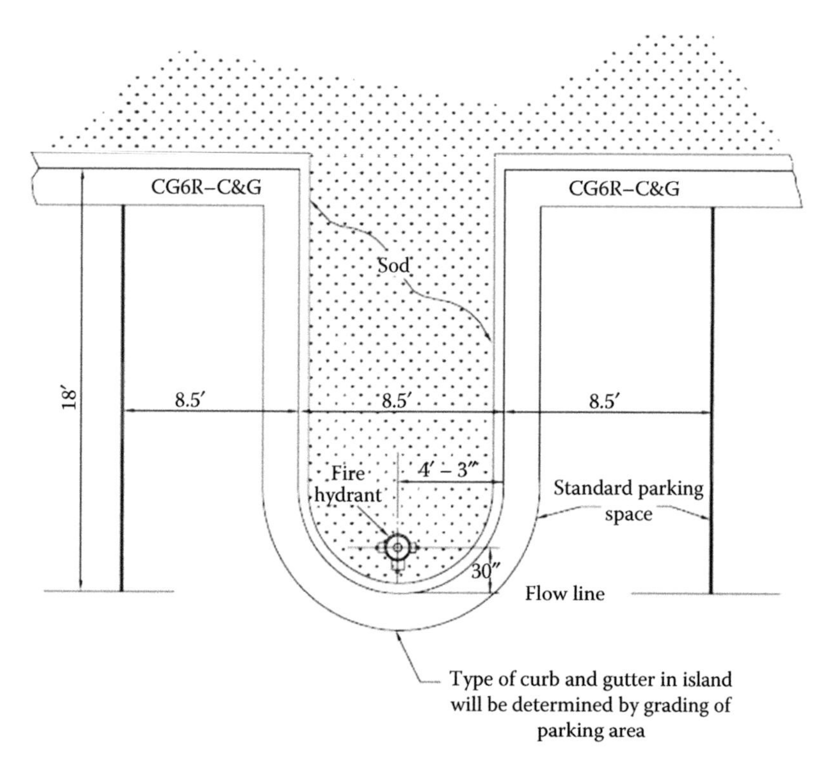

FIGURE A.5 Fire hydrant location in parking island. (Courtesy of *Fairfax County Public Facilties Manual.*)

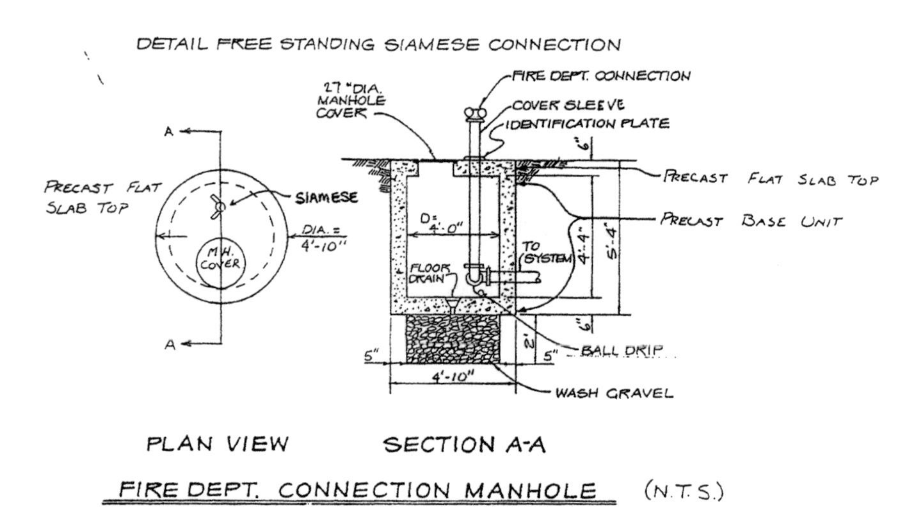

FIGURE A.6 Free-standing Siamese/fire department connection pit details. (Details provided by Fairfax County Fire Plans Review Department. With permission.)

SITE PLAN CHECKLIST

Example Fairfax County Checklist
FIRE PREVENTION DIVISION
Engineering Plans Review Section
SITE PLAN REVIEW FORM

1. Provide county site plan number or waiver number on plans.
2. Provide submitter name, address, phone number. USBC 109.5.1.
3. Provide building name, address. FXCO PFM 9-0202.2C(9).
4. State "type of construction" (IBC classification). PFM 9-0202.2C(2).
5. State "use group" (IBC classification).
6. Provide building height and number of stories. PFM 9-0202.2C(10).
7. State foot print and gross floor area of building. PFM 9-0202.2C(12).
8. If fire walls are to be built, label hour rating. PFM 9-0202.2C(11).
9. State on plans if building is to be sprinklered and show type of system to be installed. PFM 9-0202.2D.
10. Provide separate building data (items 4–9) for each building.
11. If sprinklered, show fire department connection (street front of building address side with no obstruction within 10 feet. PFM 9-0202.1J). PFM 9-0202.2C(9). Fire hydrant to be maximum 100 feet from FDC.
12. Provide location and size of underground fire lines. PFM 9-0202.2C(8).
13. Fire hydrants (existing and proposed) to be shown. PFM 9-0202.2C(4). 50 feet minimum from building and 100 feet maximum from FDC. PFM 9-0202.1F and 9-0202.1H.
14. Provide full fire flow test data—include static and residual hydrant pressure, gpm flow, available fire flow at 20 psi, vicinity map showing hydrants used for test, and connecting water line locations from test hydrants to proposed site, with elevations. PFM 9-0202.2G(1).
15. Indicate fire lanes to be painted and signage details. Make corrections as noted on Sheet(s) ___________. FXCO FPC 503.
16. Details on site plan sheets do not match those on fire lane sheet.
17. Provide fire rating of exterior walls to comply with IBC Table 602.
18. Provide adequate hydrant coverage. PFM 9-0202.1I on fire vehicle access way.
19. Indicate on plans that garage/parking deck with fire truck access and *bridges* will be designed for a nominal 450 lb/sf uniform live load.
20. Provide 20-feet-wide adequate emergency vehicle access fire lanes. FXCO FPC 503, SFPC 503 (all).
21. Include with your resubmittal of these plans the marked-up sheets from this submittal.
22. All cul-de-sacs to comply with PFM requirements (7.0406.8A0) 45 feet minimum radius.
23. When buildings are more than three stories or 30 feet in height, ladder truck access should be provided to both the front and rear of the building. The ladder truck access way shall be no less than 15 feet and no more than 30 feet from the exterior building wall.

24. Check the landscape plan for trees and other obstructions to fire trucks, ladder trucks, FDCs, and fire hydrants.
25. Must state whether garage is open or enclosed. IBC-09, 406.2.1.
26. Check water line profile. 4 feet cover minimum and 7.5 feet maximum. FCWA and PFM 9-0102.3B. Check distances to other utilities.
27. Note: 6 inch—1,500 gpm, 8 inch—2,500 gpm, 12 inch—3,500 gpm. Per FCWA. With the gpm required and a lower C value of pipe, high friction loss is experienced.
28. Freestanding FDCs to have a pit with access or a pipe to drain back to the building for ball drip.
29. 6-inch main minimum to supply fire hydrants.
30. 100 feet maximum travel fire access lane without *turnaround.*
31. Town houses need 2,500 gpm, high-rise 2,700 gpm.
32. Pools to have 12-feet-wide lanes to within 50 feet.
33. Fire hydrant must meet PFM plates. 30 inch maximum with curb. Verify fire hydrant *not* in sidewalk. PFM 9-0103. Need 3-inch cover on ditch line. PFM 9-0202.2A.
34. If building claimed to be unlimited, then check for property lines off-set and the IBC unlimited requirements.
35. Grade not to exceed 6% on fire access way or for equipment setup. NFPA 1901.
36. Check for fire wall ratings between use group for separate fire areas. IBC 706.4.

USBC = Uniform Statewide Building Code (state specific)
IBC = International Building Code (state specific)
PFM = Public Facilities Manual (county specific)
FXCO = Fairfax County
FPC = Fire Prevention Code (county specific)

Glossary

Apparatus: In the context of this text, a fire vehicle used by the fire service to perform operations.
Authority having jurisdiction (AHJ): The authority in charge of approval.
Cul-de-sac: A roadway ending in a circular turnaround.
Easement: A dedicated pathway for a utility provider to use on an owned property—a right-of-way.
Fire apparatus access: A means, such as a road, to allow fire apparatus to get to the area needed.
Fire apparatus access road: A road that provides fire apparatus access from a fire station to a facility or building, that is, fire lanes, public and private roads, parking lot lanes, and access ways.
Fire department connection (FDC): An exterior connection to the building fire suppression system used by the fire pumper to supply water and boost pressure.
Fire flow: The amount of water needed to put out the fire via manual firefighter response.
Fire separation distance: A distance, measured at right angles from the face of a building wall, to either the closest interior lot line, the center line of the street, alley, or public way, or an imaginary line between two buildings on the property.
Grade: A measurement of the angle used in road design and expressed as a percentage of elevation change over distance.
High-rise: A building with an occupied floor located more than 75 feet above the lowest level of fire department vehicle access.
Lot: An owned piece (parcel) of land separated on a map from other lots by property lines.
Residual pressure: The pressure of water when water is flowing; that is, water coming out of an open fire hydrant.
Scrub surface/area: The exterior area of the building that can be reached by an aerial ladder.
Siamese connection: See fire department connection.
Site plan: A detailed engineering drawing of proposed improvements to a given lot.
Static pressure: The pressure of a water system when water is not flowing.
Turntable: The ladder's ability to rotate 360° in the horizontal plane for aerial ladder trucks.

Index

A

Aerial ladder trucks, 64
- access, 76–81
  - enough provision of, 81–85
- effect of tree height and placement on, 79–81
- for high-rise fires, 73, 84–85
- IFC appendix recommendation, 76, 78, 81–82
- nose-in access, 77–78
- setup and rescue, 73–74

AHJ, *see* Authority Having Jurisdiction
Angle of approach, 13–14
Angle of departure, 13
Authority Having Jurisdiction (AHJ), 36, 43, 59
- for ladder truck access, 76
- for security gate access, 66
- for site plan, 1

B

Bridges, fire truck access, 66
Building address, 4–7
Building codes, 73, 83

C

"C" factor, 23
Civil scales, 7
Code modification
- existing site, 87–88
- proposed site, 87–89

Criss-cross fire hydrant, 35
Cul-de-sac, for turnaround fire truck, 68–69

D

Dead end road, 67–69
Demolition sheet, 8–9
Dry barrel, 25, 26
Dry hydrants, 25

E

Easement, fire lines, 49
ED value, *see* Exposure distance value
Energy-efficient windows, 81
Engineering scale, 7
ESFR sprinklers, xix
Existing/demolition sheet, 8–9
Exposure distance (ED) value, 20–21

F

Fairfax County method, 18–21, 36
- examples, 21–24

FDCs, *see* Fire department connections
Fire access, cases of, 68
Fire alarm notification, 9
Fire department connections (FDCs), 96
- building address and, 5, 6
- case study, 59–60
- free-standing, 58, 60, 61–62
- function of, 51–52
- for high-rise buildings, 57–59
- location, 50, 53–56
- pumper test header connection and, 60–61

Fire flow, 17
- minimum and maximum, 21, 22
- water requirement for, 23–24

Fire hydrant flow test, 23
Fire hydrants
- availability and access, 35–39
- clear space around, 31–35
- coverage, 36–38
- distance between, 32–35
- distance from curb, 95–96
- dry hydrants, 25
- fire department connections and, 51, 52, 58
- flow and pressure (Q_{20}) at, 27–31
- location, 95–96
- serving a Siamese connection, 56–57
- wet hydrants, 25–27

Fire lane, 69
- signage, 70–71

Fire lines, underground, 41–42
- case study, 49–50
- easement, 49
- NFPA 13D sprinkler systems, case study, 45–46
- profile of, 46–49
- site renovations, 49
- tall buildings, 42–45

Fire Marshal information, 1–2, 3
Fire operation, challenges for, 73
Fire protection, on site plan, xix–xxii
Fire pumper truck, 25, 26
Fire separation, 3
Fire Service Features of Buildings and Fire Protection Systems, 51
Fire site plan review, 47
Fire suppression sprinklers, xix
- failure, xx–xxii

Fire truck access, 36
  cases, 68
  characteristics of road, 63–66
  dead end road, 67–69
  fire lane signage, 70–71
Fire truck turnaround, 68–69, 93–94
Fire walls, 2–3
Flow and pressure (Q_{20}), at fire hydrants, 27–31
Freeman method, 34, 36
Free-standing FDCs, 58, 60, 61–62
Free standing Siamese connections, 96
Fuel tanks, renovation, 9

G

Grading
  applying in public *versus* private streets, 14–15
  effects on fire operations, 11–13

H

High-rise buildings, fire department connections for, 57–59
High-rise fires
  aerial ladder trucks for, 73, 84
  cases of, 84–85

I

IBC, *see* International Building Code
ICC, *see* International Code Council
IFC, *see* International Fire Code
Illinois Institute of Technology Research Institute method, 18
Insurance Service Office (ISO) method, 18, 21
International Building Code (IBC)
  and NFPA 13D sprinkler system, 45
  and NFPA 13 sprinkler system, 52
  and road access, 66, 69
International Code Council (ICC), 43
International Fire Code (IFC)
  building address and, 4, 6
  fire truck access to road, 11, 14–15, 37, 63, 65–66
  method for fire flow analysis, 21
Iowa State University (ISU) method, 17
ISO Grading Schedule method, 34
ISO method, *see* Insurance Service Office method
ISU method, *see* Iowa State University method

K

Knox box, 7

L

Low hydraulic gradient, 24

M

Maximum fire flow, 21, 22
Measuring scales, 7–8
Mechanical room building, 87
Metric scales, 7
Minimum fire flow, 21

N

National Fire Protection Association (NFPA), xix, xx
NFPA, *see* National Fire Protection Association
NFPA 14, 43
  for high-rise buildings, 4, 59
NFPA 13D sprinkler systems, 41, 83
  case study, 45–46
NFPA 1 Fire Code, 13
  and road access, 63, 65, 66, 68
NFPA 5000, for high-rise buildings, 57–58
NFPA 13R sprinkler systems, 41
NFPA 13 sprinkler systems, 24, 45–47, 52, 56, 60
NFPA 1901: Standard for Automotive Fire Apparatus, 11, 51
NFPA 24 Standard for the Installation of Private Fire Service Mains and Their Appurtenances, 27–28
  for firelines, 41, 46, 47, 50
NFPA 1142 Standard on Water Supplies for Suburban and Rural Fire Fighting, 13, 25
Nose-in access, aerial ladder trucks, 77–78

O

Occupancy reduction (OR) value, 19–20; *see also* Exposure distance (ED) value
Occupational Safety and Health Administration (OSHA), 51
OR value, *see* Occupancy reduction (OR) value

P

Pedestal/podium-style buildings, 6, 83
Plan view, fire department connections, 53–56
Profile
  of fire lines, 46–49
  of site, 8
Pumper test header, 60–61

R

Record keeping, of site plan, 8–9
Renovation, of site plan, 9
Revision, of site, 9

S

Siamese connection, 4, 7, 51
  fire hydrant serving a, 56–57
Signage, fire lane, 70–71
Site plan
  building address, 4–7
  checklist, 97–98
  damages of building due to errors in, xv–xvii
  existing condition/demolition, 8–9
  fire protection on, xix–xxii
  measuring scales, 7–8
  record keeping, 8–9, 9–10
  review, 1–4
Site renovations, 9
  fire lines, 49

T

Tall buildings, fire lines in, 42–45

U

Uniform Fire Code, 63, 65
U.S. Fire Administration (USFA), xix

V

Vehicle blocking access to fire hydrant, 33

W

Water
  in fire line, 47
  flow for fire hydrants, 28–30
Water tanks, for fire flow, 23
Wet hydrants, 25–27

For Product Safety Concerns and Information please contact our EU representative GPSR@taylorandfrancis.com Taylor & Francis Verlag GmbH, Kaufingerstraße 24, 80331 München, Germany

T - #0129 - 230425 - C0 - 234/156/6 [8] - CB - 9781498741781 - Gloss Lamination